The Science
of
Marijuana

The Science of Marijuana

Leslie L. Iversen

OXFORD
UNIVERSITY PRESS

2000

OXFORD

UNIVERSITY PRESS

Oxford   New York
Athens   Auckland   Bangkok   Bogotá   Buenos Aires   Calcutta
Cape Town   Chennai   Dares Salaam   Delhi   Florence   Hong Kong   Istanbul
Karachi   Kuala Lumpur   Madrid   Melbourne   Mexico City   Mumbai
Nairobi   Paris   São Paulo   Singapore   Taipei   Tokyo   Toronto   Warsaw

and associated companies in
Berlin   Ibadan

Copyright © 2000 by Oxford University Press, Inc.

Published by Oxford University Press, Inc.,
198 Madison Avenue, New York, New York, 10016
http://www.oup-usa.org

Oxford is a registered trademark of Oxford University Press

Library of Congress Cataloging-in-Publication Data
Iversen, Leslie L.
The science of marijuana / Leslie L. Iversen.
p.   cm.   Includes bibliographical references and index.
ISBN 0-19-513123-1
1. Marijuana — Physiological effect.   2. Marijuana — Toxicology.
I. Title.   [DNLM: 1. Tetrahydrocannabinol — pharmacology.
2. Cannabis — adverse effects.   3. Cannabis — therapeutic use.
4. Central Nervous System — drug effects.
QV 77.7 I94s 2000]   QP801.C27I94   2000
615'.7827 — dc21   DNLM/DLC   for Library of Congress   99-32747

2 4 6 7 9 7 5 3 1
Printed in the United States of America
on acid-free paper

Foreword

SOLOMON H. SNYDER

More than virtually any drug in history, cannabis exemplifies the adage that history repeats itself again and again — and we never learn. My first focus on cannabis came in 1970 when I wrote a book on it for the general public. During the relatively brief interval between then and now, thinking has veered several times in diverse directions, especially in the United States, clearly justifying the fresh look provided by the present volume.

The late 1960s witnessed a renaissance of cannabis research following Raphael Mechoulam's discovery that the active ingredient in cannabis is delta-1-tetrahydrocannabinol (THC). There was a flurry of excitement by drug companies attempting to develop derivatives of THC that might exert therapeutic effects but would not be psychoactive. In the sociopolitical realm, the era's youth culture and freethinking about drugs led to progressively "softer" cultural views regarding the use of cannabis for recreational purposes. Over the next two decades the inevitable retrogression ensued. Researchers failed to identify safe and effective derivatives of THC. Social and political attitudes in the United States and Western Europe became progressively more conservative with police pursuing users and dealers with ever greater fervor.

In the mid 1990s attitudes changed once more with a focus on medical uses. Though the pharmaceutical industry still had not developed effective THC derivatives, more and more studies were demonstrating that smoking cannabis enables patients with advanced cancer to cope with pain, eases the nausea caused by chemotherapeutic drugs used to treat cancer, and lowers intraocular pressure in patients with glaucoma. Several states passed laws permitting the medical use of cannabis followed shortly by concerns that recreational users might illicitly feign medical needs. The House of Lords in the United Kingdom carried out a study weighing the good and bad of cannabis, while the Institute of Medicine of the United States National Academy of Sciences carried out a similar investigation. Both came to the conclusion that cannabis has both good and bad points and advised the general public and law-making bodies to adopt a balanced perspective. While smoking cannabis for recreational purposes may indeed cause some harm, it is not nearly as dangerous as cocaine, alcohol, or cigarettes. As for medical uses, some areas are quite promising, but more clinical studies are needed.

What I find remarkable about all of this is that the British and American Commissions, reflecting recent public thinking, bear remarkable similarities to analogous efforts going back at least 100 years. In the nineteenth century the British government addressed the uses of cannabis in India, where it had been employed routinely for recreational and medical purposes since the first millennium. The 1893 report of the Indian Hemp Drugs Commission is particularly impressive in its "modern," well-balanced perspective. At that time cannabis was rumored to elicit criminal behavior and severely damage the brain. The Commission critically evaluated all reports and concluded, like the United States and the United Kingdom reports 100 years later, that cannabis had good as well as bad features, and therapeutic promise as well as recreational utility if employed in moderation.

There were further incarnations of the cannabis debate. In the United States in the late 1930s and early 1940s researchers had come close to isolating the active ingredient of the plant and had synthesized THC derivatives that were entering clinical trial. However, many news media reports labeled cannabis a "killer drug." In 1939, Mayor LaGuardia of New York City appointed a Blue Ribbon Committee to evaluate the situation that came to conclusions similar to those of the Indian Hemp Commission. Especially notable are comments of the Committee dealing with allegations that cannabis is a stepping stone to the use of more dangerous drugs, something that we hear again and again in the 1990s. The report concluded, "the use of marijuana does not lead to morphine or heroin or cocaine addiction, and no effort is made to create a market for these narcotics by stimulating the practice of marijuana smoking."

Professor Iversen served as a consultant to the House of Lords Committee on cannabis, which prompted him to conduct a thorough review of the literature leading to the present volume. He has thoughtfully reviewed the history of cannabis so that one can appreciate current concerns based on their historical context. He has carefully evaluated numerous studies dealing with uses of cannabis in medicine, and he analyzed the various pharmacological actions of the drug and possible toxic actions in different organs. He has presented his sophisticated analyses in a lucid and entertaining mode.

Many books have been written about marijuana and I have read almost all of them. Iversen's volume is far and away the most scholarly, elegantly assembled, and clearly presented. Read on.

Preface

As a scientist who works on understanding how drugs act on the brain, I am exasperated by the way in which the proponents and opponents of cannabis use and abuse science in defending their positions. This is a drug whose actions have been studied in some detail; there is a considerable scientific literature on how it acts and the possible adverse effects associated with long-term use. There have also been exciting new scientific advances in the past few years with the discovery that the brain contains its own cannabis-like chemical messenger system — a finding potentially as important as the much publicized discovery of a naturally occurring series of morphine-like chemicals in the brain — the endorphins — in the 1970s. There have also been claims that cannabis has important medical applications and these too have been researched — although less thoroughly so far.

Cannabis has been demonized, however, in public debate and the available scientific information is largely ignored or distorted by various groups who use science as a propaganda weapon. In the book *Marijuana Myths and Marijuana Facts* published in 1997, the authors, L. Zimmer & J.P. Morgan, say that *"Marijuana's therapeutic uses are well documented in the modern scientific literature,"* while in July of the same year the British Minister of Health in reply to a parliamentary question about the medical uses of cannabis said, *"At present the evidence is inconclusive. The key point is that a cannabis-based medicine has not been scientifically demonstrated to be safe, efficacious and of suitable quality,"* and in August 1996, General Barry McCaffrey, the United States drug czar somewhat more bluntly said, *"There is not a shred of scientific evidence that shows that smoked marijuana is useful or needed. This is not medicine. This is a cruel hoax."*

In an unprecedented move, voters in a number of states in the United States approved proposals in the1996 and 1998 elections to approve the medical use of marijuana, although these laws have so far not been fully implemented. While marijuana is available for personal use in Holland, and its decriminalization is being considered in Denmark, Switzerland, and Greece, official government policy in the United States and in most European countries remains firmly against any change in the present laws that treat marijuana as a scheduled drug with no medical uses whose possession is a criminal offense. These laws are vigorously enforced, with more than half a million marijuana arrests every year in the United States.

This book seeks to describe what is known about how marijuana acts in the brain, and to compare the profile of marijuana with other drugs that are used because of their euphoriant or psychostimulant effects — cocaine, amphetamines, heroin, alcohol, and nicotine. Is marijuana, like these, a drug of addiction with harmful side effects — or is it a "soft drug" the use of which is not harmful to health? Does it have genuine medical uses that cannot be filled by other existing medications? No drug is without adverse effects–several thousand people die every year from the adverse effects of aspirin and related painkillers. Cannabis too has adverse effects — but how serious are they, and do they justify prohibition of the legal use of the drug? The present review may help to inform the debate on public policy that is currently under way. I am not sanguine about the outcome of this debate in the immediate future — smoked marijuana seems to carry with it too many potential long-term risks to health to justify its widespread use, while other formulations of the drug for medical applications leave much to be desired. Attitudes against cannabis are also deeply entrenched. In the longer term, the development of better technologies for delivering the active principle of marijuana medicinally and the discovery of completely new therapeutic strategies as a result of new advances in basic research on cannabinoid systems in the body may hold the greatest promise.

L.I.

Oxford, U.K.
April 1999

Contents

The Science
of
Marijuana

1

Introduction

M arijuana (cannabis) is among the most widely used of all psychoactive drugs. Despite the fact that its possession and use is illegal in most countries, cannabis is used regularly by as many as 20 million people in the United States and Europe, and by millions more in other parts of the world. In recent years thousands of patients with acquired immunodeficiency syndrome (AIDS), multiple sclerosis, and a variety of other disabling diseases have begun to smoke marijuana illegally in the firm belief that it makes their symptoms better, despite the relative paucity of medical evidence to substantiate such belief.

The writer Stephen Jay Gould described his use of marijuana in suppressing the nausea associated with cancer treatment:

> I had surgery, followed by a month of radiation, chemotherapy, more surgery, and a subsequent year of additional chemotherapy. I found that I could control the less severe nausea of radiation by conventional medicines. But when I started intravenous chemotherapy (Adriamycin), absolutely nothing in the available arsenal of antiemetics worked at all. I was miserable and came to dread the frequent treatments with an almost perverse intensity.

> . . . marijuana worked like a charm. I disliked the " side effect" of mental blurring (the "main effect" for recreational users), but the sheer bliss of not experiencing nausea — and then not having to fear it for all the days intervening between treatments — was the greatest boost I received in all my year of treatment, and surely had a most important effect upon my eventual cure.

> (Grinspoon and Bakalar, 1993)

In California as part of the 1996 election voters approved "Proposition 215," which sought to make it legal to smoke marijuana with a doctor's recommendation. During the following year "cannabis buyers clubs" were established throughout the state to provide supplies of cannabis for medicinal use. On the whole these were run by well-intentioned people and were strictly regulated, patients were checked for identity, medical records, and doctor's diagnosis and only then were they allowed to purchase a small quantity of marijuana. In the 1998 state elections a further six states voted to permit access to medical marijuana, and in early 1999, such laws began to be put into effect, despite opposition from the federal government.

In Amsterdam, the Blue Velvet Coffee Shop is on a busy city street adjacent to shops and cafes. Inside it seems to be a small, friendly and ordinary place, one of more than 2000 similar establishments in Dutch cities. There are a few posters on the wall, a coin-operated video game, and loud music. Behind the bar along with the usual espresso machine and soft drinks, the menu features 30 varieties of cannabis resin and 28 varieties of marijuana leaf. Customers come in to purchase a small *bag* or some *hash brownies* to take away, and some linger to smoke marijuana cigarettes on the premises with their cappuccino. Regular customers have their "loyalty card" stamped with each purchase (one bag free as a bonus for every four purchased).

In the autumn of 1997 the respectable British newspaper *The Independent on Sunday* launched a campaign in favor of the decriminalization of cannabis. The campaign attracted thousands of supporters from the medical profession and all other walks of life. For the first time for more than 25 years in Britain the issue was one of public debate.

Does all this mean that Western society is starting to take a more liberal view toward cannabis use, one that tends toward the Dutch assessment of it as a "soft drug" that should be distinguished and separated from "hard drugs?" Far from it. The Californian cannabis buyers clubs were closed in 1998 after intense pressure from the federal government, following a state appellate court judgement that they were illegal, notwithstanding Proposition 215. The United States federal government, furthermore, threatened to punish any Californian doctors willing to recommend the use of marijuana to their patients. In Arizona, where voters also approved the medical use of marijuana, the state government rapidly put a stop to any moves to implement this after the 1996 election. Even in liberal Holland, the coffee shops have no legal means of obtaining their supplies of cannabis, and in 1997, the Dutch government, under international pressure, felt obliged to close down the buyers club operation that had been set up specifically to provide cannabis for medical use.

In London during the Christmas holiday period of 1997, William Straw, 17-year-old son of the Cabinet Minister Jack Straw, was befriended in a London pub by two attractive women who plied him with beer and compliments. At the end of the evening he was only too happy to acquiesce to their request by purchasing a small quantity of marijuana for them from the local dealer. The women, however, turned out to be journalists from the tabloid newspaper the *Daily Mirror* and William had been set up. After the journalists told their story he was arrested for possible drug dealing and for a week after this the British newspapers had a field day in their subtle and not so subtle attempts to reveal which cabinet minister's son had been arrested (his name could not legally be disclosed as William was under age). In the event William suffered nothing more severe than a police warning, and his embarrassed father felt obliged to explain to the British public that cannabis was a dangerous narcotic drug, adding for good measure that it had no proven medical utility. Other young people caught in possession or selling small quantities of cannabis to their friends are often not so lucky, losing their place at school or university and not infrequently ending up with a prison sentence.

Who is right? Is cannabis a relatively harmless "soft drug?" Does it have genuine medical uses that cannot be fulfilled by other medicines? Or, is the campaign to legalize the medical use of cannabis merely a smokescreen used by those seeking the wider acceptance of the drug? Is cannabis in fact an addictive narcotic drug that governments are right to protect the public from? This book reviews the scientific and medical evidence on cannabis, and tries to answer some of these questions. Often in analyzing the mass of scientific data it is difficult to come to clear-cut conclusions. To make matters worse in this particular case, the opposing factions in the cannabis debate often interpret the same scientific evidence differently to suit their own purposes.

This introductory chapter introduces the hemp plant from which the various cannabis products derive and gives a brief history of the drug

The Plant

The hemp plant (*Cannabis sativa*) probably originated in Central Asia but through man's activities, has been distributed widely around the

world (for a comprehensive review of cannabis botany see Clarke, 1981)
It has been cultivated as a multipurpose economic plant for thousands of
years and through the process of selection for various desirable characteristics, many different cultivated varieties exist—some grown exclu-
sively for their fiber content, others for their content of psychoactive
chemicals. All of these varieties, however, are generally classified as a
single species first named in 1735, by the famous Swedish botanist Lin-
naeus, as *Cannabis sativa*. The Cannabis plant is a lush, fast-growing
annual, which can reach maturity in 60 days when grown indoors under
optimum heat and light conditions, and after 3–5 months in outdoor
cultivation. The plant has characteristic finely branched leaves subdi-
vided into lance shaped leaflets with a saw tooth edge. The woody, angu-
lar, hairy stem may reach a height of 15 feet or more under optimum
conditions. A smaller more bushy subspecies reaching only 4 feet or so in
height known as *Cannabis indica* was first described by Lamark and is
recognized by some modern botanists. There is currently much activity
among plant breeders in Holland (where cultivation of the plant for personal use is legal) and in California (where such cultivation is illegal) to
produce new varieties with increased yields of the psychoactive chemical
delta-9-tetrahydrocannabinol (THC). The details of the breeding pro-
grams are not public, but involve such techniques as the treatment of
cannabis seed with the chemical colchicine to cause the creation of poly-
ploid plants, in which each cell contains multiple sets of chromosomes
instead of the normal single set. Such varieties may have extra vigor and
an enhanced production of THC, although they tend to be genetically
unstable. Other varieties have been obtained by crossing *Cannabis sativa*
with *Cannabis indica* strains, to yield a number of different hybrids.
These strains may not breed true but by selecting the first generation (F1
hybrid) seeds of such crosses, plants can be generated with hybrid vigor
and enhanced THC production. Particularly favorable genetic strains
can also be propagated vegetatively by cuttings—in this way a single
plant can give rise to thousands of clones with identical genetic makeup
to the original. Although the cultivation of cannabis for THC production
is illegal in most Western countries, the Internet carries advertisements
from numerous seed companies who offer to supply seeds of as many as

30 different varieties of cannabis — with such names as "Skunk," "Northern Lights," "Amstel Gold," and "Early Girl." Prices for individual seeds average $2–4 — but in something approaching the seventeenth century "tulipmania" Dutch suppliers seek as much as $15 for a single seed of "White Widow," the winner of the "High Times Cup" in the annual Amsterdam Cannabis Festival in 1995 and 1996. With the power of modern plant breeding techniques and the possibility of genetic engineering to enhance THC production there seems little doubt that even more potent varieties of *Cannabis sativa* can and probably will be created.

The cannabis plant is either male or female and under normal growing conditions these are generated in roughly equal numbers. The male plant produces an obvious flower head, which produces pollen, while the female flower heads are less obvious, and contain the ovaries ensheathed in green bracts and hairs (Fig. 1.1). The psychoactive chemical THC is present in most parts of the plant, including the leaves and flowers, but it is most highly concentrated in fine droplets of sticky resin produced by glands at the base of the fine hairs that coat the leaves and particularly the bracts of the female flower head. The resin may act as a natural varnish, coating the leaves and flowers to protect them from desiccation in the hot, dry conditions in which the plant often grows. Contrary to the ancient belief that only the female plant produces THC, it is now clear that the leaves of male and female plants contain approximately the same amounts of THC, although the male plant lacks the highly concentrated THC content associated with the female flowers. If pollinated, the female flower head will develop seeds that contain no THC but have a high nutritional value. Indeed cannabis was an important food crop — listed as one of the five major grains in ancient China — and is still cultivated for this purpose in some parts of the world today. From the point of view of the cannabis smoker, however, the presence of seeds is undesirable: they burn with an acrid smoke and tend to explode on heating, and their presence dilutes the THC content of the female flower head. In the cultivation of cannabis for drug use in India it was customary to remove all the male plants from the crop as they began to flower to yield the resin-rich sterile female flowering heads, which were then dried and compressed to form the potent product known as *ganja*.

The services of expert *ganja doctors* were often employed, who went through the hemp field with an expert eye cutting down all the male plants before they could flower. The labor intensive process of removing all male plants is also used today in some parts of the West to produce sterile female flower heads, known as *sensemilla*. These may contain up to five times more THC than the *marijuana*[1] produced from the dried leaves of other parts of the plant (Table 1.1). The most potent preparation derived directly from the plant is *hashish*, which represents the THC-rich cannabis resin obtained by scraping the resin from the flower heads, or by rubbing the dried flower heads and leaves through a series of sieves to obtain the dried particles of resin — known as *pollen*. These are compressed to form a cake of yellow to dark brown hashish. A more colorful method of obtaining the pure resin in India was described in 1840 by the Irish doctor William B. O'Shaugnessy, who worked for many years in India:

> Men clad in leather dresses run through the hemp field, brushing through the plants with all possible violence; the soft resin adheres to the leather, is subsequently scraped off and kneaded into balls, which sell from five to six rupees the seer. A still finer kind . . . is collected by hand in Nepal — the leather attire is dispensed with, and the resin is gathered on the skins of naked coolies.

(O'Shaugnessey, 1842)

Another more recently popular product is *cannabis oil* produced by repeatedly extracting hashish resin with alcohol. The concentrated alcoholic extract may vary in color from green (if prepared from resin containing significant amounts of fresh cannabis leaf) to yellow or colorless for the purer preparations. It can contain up to an alarmingly high 60% THC content, but more usually the THC content is around 20%. Nevertheless, one drop of such oil can contain as much THC as a single marijuana cigarette.

The Cannabis plant develops in many different ways, according to

---

1. The term *marijuana* is used widely in North America to describe herbal cannabis — in Europe the word *cannabis* is more common. In this book the two words will be used interchangeably.

A

Figure 1.1. Engravings showing the characteristic appearance of the flowering heads of female *(A). (Continued)*

B

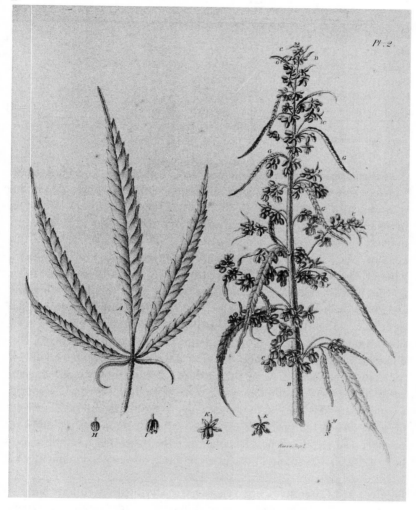

Figure 1.1. (*Continued*). Male *(B)* cannabis plants. From Wisset (1808).

Table 1.1. Cannabis Preparations

Name	Part of Plant	THC Content (%)
Marijuana (bhang, dagga, kif)	Leaves, small stems	1.0–3.0
Sensemilla	Sterile female flower heads	3.0–6.0
Ganja	Compressed sterile female flower heads	4.0–8.0
Hashish (charas)	Cannabis resin	10.0–15.0[a]
Cannabis oil	Alcoholic extract of resin	20.0–60.0

[a]Street samples of cannabis resin often contain much smaller quantities of THC because they are frequently adulterated with other substances.

the genetic variety and the soil, temperature, and lighting conditions under which it is grown. To generate optimum quantities of THC-containing resin the plant needs a fertile soil and long hours of daylight, preferably in a sunny and warm climate. This means essentially that for THC production outdoor growth occurs optimally anywhere within 35° of the equator. Typical growing regions include Mexico, northern India, many parts of Africa, Afghanistan, and California. In northern Europe and in Russia the plant has long been cultivated as a fiber crop but such plants are grown from varieties selected for this purpose and do not generate significant amounts of THC. Nowadays much culture of cannabis takes place indoors (Fig. 1.2), where nutrients, lighting, and temperature conditions can be optimized, and the cultivation (illegal in most countries) more easily concealed. More than half of the cannabis consumed in the coffee shops in the Netherlands is grown domestically indoors. The large variability in THC production according to strain and culture conditions presents one of the problems associated with its use as a drug either for medical or recreational use. In either case the consumer has little indication of the THC content of the plant product, and may consequently fail

Figure 1.2. Dr. Geoffrey Guy, Chairman of GW Pharmaceuticals Ltd., examines *cannabis sativa* plants at the company's climate-controlled greenhouse, somewhere in Southern England. The British government has licensed the cultivation of these plants to permit clinical trials of cannabis extracts for various medical indications. Photo courtesy of GW Pharmaceuticals Ltd.

to obtain an adequate dose or alternatively may unwittingly take a larger dose than desired.

The cannabis plant is currently thought of mainly in the context of the psychoactive drug THC, but it is a versatile species that has had a very important place in human agriculture for thousands of years (for review see Robinson, 1996). An acre of hemp produces more cellulose than an acre of trees, and the tough fiber produced from the outer layers of the stem has had many important uses. Hemp fiber made the ropes that lifted up the tough hemp-derived canvas cloth (the word derives from the Dutch pronunciation of cannabis) used to make the sails of the ancient Phoenician, Greek, and Roman navies. Archaological evidence shows that hemp fiber production was going on in north eastern Asia in Neolithic times, around 600 B.C, and hemp production spread around the world, including to the United States where it was introduced by the first settlers. Although the importance of hemp declined with years, there were still 42,000 acres cultivated in the United States in 1917. Other major commercial centers of production were in Europe and in Russia. Ship sails, ropes, clothing, towels, and paper were all derived from hemp fiber and the woody cellulose-rich interior *hurds* of the hemp stem. Until the 1880s almost all of the world's paper was made from hemp, and even today many bank notes are still printed on cannabis paper because of its toughness and durability. Most of our great art works are painted on canvas, and the first jeans were made from canvas cloth.

Robert Wissett in his *Treatise on Hemp* (1808) gave a comprehensive account of the cultivation of hemp as a fiber plant in Europe, Asia, and the United States 2 centuries ago. The method of cultivation as practiced at Crowland in Lincolnshire in England in the early nineteenth century was described:

Hemp should be sown about the last week in April. It requires a good soil, and will not thrive in clays or cold stiff lands. To produce Hemp the land should be plentifully manured, in the proportion of twenty-two loads per acre. The manure is spread and ploughed in a short time previous to the sowing. In this country one ploughing is thought sufficient. Three bushels of seed are generally allowed for an acre. The land should be cleared of weeds before sowing. It seldom happens that any further weeding is requisite; if weeds do appear, the Hemp itself soon chokes them. About the end of September Hemp ripens; it is then pulled up by the roots, and tied into sheaves, of the size of ordinary corn sheaves. Wages for pulling are, upon average, about a shilling (5p/US$0.08) per score of sheaves. In a few days the sheaves are formed into shocks each of which consists of one hundred sheaves. A cloth is laid between every three sheaves for the convenience of threshing, and to receive the seed which may casually fall out. The shocks are covered with Hemp-lop, i.e. barren and withered stalks, to protect them from the weather, birds, etc. In this state they stand for about three weeks or a month; should the weather prove wet, a longer time will perhaps be necessary.

The seed is then threshed out in the field, into the cloths, which were before stated to be placed between every three sheaves. After threshing, the Hemp is covered close with sods in stagnant water. Care must be taken to exclude all fresh water after the immersion of the Hemp, otherwise the tendency to peel, which is the intent of this process, would then be delayed. After having been thus steeped about three weeks, Hemp is usually fit to peel; it is then placed in the fields for about a week (in fine weather) to dry: afterwards removed under shelter, and peeled by women. The price of the labour is about seven pence (US$0.10) per stone (14 lb). After peeling the stalks are formed into bundles of the size of a common faggot, and sold for one penny-per bundle as fuel, which purpose they answer extremely well. There is also another way of making Hemp, called breaking, which is performed by a machine named a Hemp-break; this method is but little used at Crowland, except for the small stalks, which it would be tedious to peel. The Hemp by breaking is rendered finer and more fit for the manufacture of linen; for this purpose, however, it should be pulled before it ripens, and thus the profit arising from the seed is lost.

(Wisset, 1808)

The author goes on to detail the economics of hemp cultivation. For each acre a clear profit of £8 ($13.00) was reported, after expenses of £14.70 ($23.22). Despite the labor-intensive methods used, the total cost of labor was a mere £3.40 ($5.40) per acre.

The hemp seed has also been an important food crop, and from it can be derived an oil that has many uses as a lubricant, paint ingredient, ink solvent, and cooking oil. The seeds are now used mainly for animal feed and as birdseed.

Most of the ancient uses of hemp have been overtaken by the ad-

vent of cotton goods, synthetic fibers, forestry-derived paper, and alternative food grains. Nevertheless, the cultivation of hemp as a fiber crop still
continues on a small scale in Europe with the sanction of the European
Union; farms in Hampshire in southern England, a traditional center for
hemp farming, continue to grow the crop. At the present time, some
lobby groups are seeking with almost evangelical fervor to increase the
cultivation of hemp on environmental grounds as a valuable biomass
crop from which fuels can be derived, or from which building materials
and other industrial products can be derived (Robinson, 1996; Herer,
1993, and web sites: *The Cannabis Shipping Company*, **www.cannabis/
shipping.co.uk**; *Institute for Hemp*, **www.hemp.org/hemp.html**; *Hemp
Union (UK) Ltd*, **www.karoo.net/hemp-union/main.htm**).

Consumption of Cannabis Preparations for Their Psychoactive Effects

Cannabis products have been consumed for thousands of years in different human cultures. It is not surprising that this has taken many different
forms, some of the more common are described here.

Smoking

Smoking is one of the most efficient ways of ingesting cannabis and
rapidly experiencing its effects on the brain (see Chapter 3). The favorite
of many people in the West is the marijuana cigarette. This consists of a
variable quantity of dried marijuana leaf (from which stems and seeds
have first carefully been removed), rolled inside a rice paper cylinder
either by hand or using a rolling machine. A typical marijuana cigarette
would contain between a half to one gram of leaf with or without added
tobacco — which assists the otherwise often erratic burning of the marijuana. A modern version consists of a cigar from which the tobacco filling has been removed and replaced with herbal marijuana. Many different slang words describe herbal marijuana, e.g., "Aunt Mary," "Dope,"
"Grass," "Joint," "Mary Jane," "Reefer," "Spliff," and "Weed." When a

joint has been smoked down to the point that it is difficult to hold it is called a *roach*, and this still contains appreciable amounts of THC that gradually distil down the length of the cigarette as it is smoked. The roach may be held in the split end of a match or with a wide variety of *roach pins* with which one may hold the roach without burning oneself. In the social groups in which marijuana is commonly smoked, as with the port served in Oxford and Cambridge Colleges after dinner, etiquette demands that the joint is passed around the group in a circular fashion. As with the port, hoarding of the joint by any one person is regarded as a serious breach of protocol. Experienced marijuana smokers often develop the technique of inhaling a considerable quantity of air along with the smoke — this dilutes the smoke making it less irritating to the airways and allowing deeper inhalation. Marijuana smokers tend to inhale more deeply than cigarette smokers do and to hold the air in their lungs for longer before exhaling.

Marijuana can also be smoked using a variety of pipes. A simple pipe resembling those used for tobacco can be used, but marijuana pipes are made of such heat-resistant materials as stone, glass, ivory, or metal. This is necessary because marijuana does not tend to stay alight in a pipe so it constantly has to be relit. A common variety of pipe is the water pipe or bong. These come in many different forms but all use the same principle. Smoke from the pipe is sucked through a layer of water, which cools it and removes much of the tar and other irritant materials present in marijuana leaf smoke. Bongs tend to be complex and heavy devices and thus not easily portable.

The more potent forms of cannabis, sensemilla, ganja, and cannabis resin are also often smoked using cigarettes or a pipe, and commonly mixed with tobacco. Pipe smoking is the traditional method for smoking ganja in India and hashish in the Arab world. Khwaja A. Hasan gives the following description of ganja smoking in contemporary India:

> Ganja is smoked in a funnel-shaped clay pipe called chilam. Almost anybody except the untouchables (sweeper caste) can join the group and enjoy a few puffs. The base part of the bowl portion of the funnel-shaped pipe is first covered with a small charred clay filter. Then the mixture of ganja and tobacco is placed on this filter. A small ring, the size of the bowl,

of rope fibre called baand is first burnt separately and then quickly placed on top of the smoking material. The pipe is now ready for smoking. Usually four or five people gather around a pipe . . . ritual purity of the pipe is always preserved for the clay pipe is never touched by the lips of the smoker. The tubular part of the chilam at its bottom is held in the right hand and the left hand also supports it. The passage between the index finger and the thumb of the right hand is used in taking puffs from the pipe . . .while they sit in a squatting position on a chabootra (raised platform) in front of one person's house, or gather in an open space while the host prepares the chilam they talk about social problems, weather, crops, prices, marriage negotiations and so forth. Such gatherings may take place at any time during the day except early morning. After a smoke they again go back to work. Thus such smoking parties are like "coffee breaks" in the American culture.

(Hasan, see Rubin, 1975)

Other observers, however, have not viewed the communal smoking of cannabis in such a benign light. Dr. Pablo Osvaldo Wolff, a member of the World Health Organization (WHO) Expert Committee on Habit Forming Drugs was alarmed by the spread of marijuana in Latin America. In 1949, he wrote:

There are many in Brazil, Mexico and Cuba who prefer to smoke marihuana collectively, but parties of this kind are not friendly gatherings but rendezvous of the vicious. The marihuana fiend can, therefore, be classified as a "gregarious addict." . . .The meetings of such "diambista" clubs are generally held on Saturdays, at the home of the eldest of the participants or of the one who has greatest influence among them. Very quickly a scene is presented typical of the old-time madhouses: men in a complete state of intoxication, delirious hilarity, with all the intermediary stages, flights and pursuits, cries and uproar, indecent songs and bawdy verses, always dedicated to the drug and in which African words are intermingled. Some already in a furious state or an aggressive attitude, become dangerous; others in a state of prostration, languish or exhausted, sleep profoundly.

(Wolff, 1949)

In modern Western society the use of cannabis oil has been introduced (a very potent alcoholic extract of cannabis resin). A few drops of this in a normal tobacco cigarette offers a means of smoking cannabis that is hard to detect. In some parts of the United States cocaine freebase (crack) and occasionally cocaine hydrochloride are mixed with marijuana and smoked. A marijuana cigarette laced with crack cocaine is commonly called a *grimmie*. Although smokers of grimmies are smoking crack cocaine, they seem to view it as a less intense and less severe form than smoking cocaine from a pipe.

Eating and Drinking

Tetrahydrocannabinol is soluble in fats and in alcohol so it can be extracted and added to various foodstuffs and drinks and taken into the body in that way. This method of consumption gives a much slower absorption (see Chapter 2) and avoids the irritant effects of inhaled smoke that many people find objectionable. The heating of marijuana during the preparation of foodstuffs or drinks leads to the formation of additional THC from the chemical breakdown of inactive carboxylic acid THC derivatives present in the plant preparations. A common method is to heat the plant leaf in butter, margarine, or cooking oil and then to strain out the solid plant materials and to use the oil or butter for cooking— often to make cakes and biscuits (e.g., hash brownies). Tetrahydrocannabinol can also be extracted with alcohol by heating and straining, yielding a variety of tinctures (e.g., green dragon), which can be diluted with lemonade or other flavored drinks. In the United States and British medical use of cannabis, the formulations used were alcoholic extracts of the plant, sometimes diluted further with alcohol to yield "Tincture of cannabis." These were diluted with water and administered by mouth.

In India, smoking marijuana in the form of cigarettes has never been popular. *Bhang* (marijuana) is commonly rolled into small balls and eaten, or infused in boiling water with or without added milk to form a drink. Such methods yield preparations with only modest amounts of THC—as the active compound is not water-soluble. The fats present in milk, however, make this a more effective means of extracting THC. In Indian cities bhang is sometimes added to the milk used for making an ice cream called *gulfi*. Many different cannabis-containing drinks and foods are known in Indian culture. Khwaja A. Hasan gives the following description of the famous decoction prepared from bhang called *thandai*.

> Preparing thandai is a time-consuming process. A number of dry fruits, condiments and spices are used in its preparation. Almonds, pistachio, rose petals, black pepper, aniseed, and cloves are ground on the toothed grinding plate (silauti); water is added so that a thinly ground paste is obtained. This paste is dissolved in milk and then bhang is added to the mixture. A few spoons of sugar or jaggery (boiled brown sugar) are added finally and then the decoction is ready for consumption. . . . The preparation of thandai and the social atmosphere it creates has great significance. Members of the same family, caste or a circle of friends from the village or the neighbourhood gather in the parlour of a friend. Different ingredients of the drink are collected and ground on the toothed stone grinding plate. The whole process takes an hour or so. While preparing the drink, individuals talk about friends, family members, prices of goods and services and a host of other problems.
>
> (Hasan, see Rubin, 1975. Reprinted with permission from
> Monton de Gruyter, Berlin)

Around the world, a variety of different cannabis preparations have been devised in different cultures and the diversity of this range equals the many different forms in which human beings have traditionally consumed alcohol — from light beer to distilled spirits, from vin de table to Premier Cru Chateau-bottled clarets.

A Brief History

Excellent reviews of the long history of cannabis can be found in Abel (1943), Lewin (1931), Robinson (1996), and Walton (1938). Evidence of man's first use of cannabis has been found in fragments of pottery bearing the imprint of a cord-like material thought to be hemp, in China, dating around 10,000 B.C. Other early evidence for hemp cultivation comes from the finding of fragments of hemp cloth in Chinese burial chambers from the Chou dynasty (1122–265 B.C.). It seems likely that hemp was cultivated and used for the manufacture of ropes, nets, canvas sails, and cloths in ancient China. The first descriptions of the medical and intoxicant properties of the plant are to be found in the ancient Chinese herbal, Pen-ts'ao, ca. 1–2 century AD. Classical myth relates that the Chinese deity Shen Nung tested hundreds of herbal materials in a series of heroic experiments in self-medication and agronomics. So potent was this myth of the etiology of medicine that the god's name was attached to the Pen-ts'ao. This herbal pharmacopoeia describes hundreds of drugs, among them cannabis which was called *ma*, a pun for "chaotic." This ancient text clearly describes the stupefying and hallucinogenic properties of the plant. Pharmacologists and herbalists added sections to the text for many centuries and Chinese physicians used cumulative editions of Pen-ts'ao as the standard text on medical drugs for hundreds of years. Shen Nung, the Farmer God, became the patron deity of medicine, with the title "Father of Chinese Medicine." Ma, often mixed with wine in a preparation called *ma-yao*, was used principally for its pain-relieving properties. Although there seems also to have been some use of the drug as an intoxicant in China, this never became widespread.

In contrast to China, the use of cannabis for its psychoactive properties has been endemic in India for more than a thousand years. Cannabis use was known by the nomadic tribes of north-eastern Asia in Neolithic times, and may have played an important role in the practice of the religion of shamanism by these people. The nomads brought the plant and its uses to western Asia and then to India. Ancient Indian legend tells how the Hindu god Siva became angry after a family row and wandered off into the fields by himself. Exhausted by the heat of the sun he sought shade and refuge under a leafy plant and finally went to sleep. On waking he became curious about the plant that had given him shelter and ate some of its leaves. This made him feel so refreshed that he adopted it as his favorite food. From then on Siva was known as the Lord of bhang (the Indian term for marijuana). In ancient Indian texts, bhang is referred to in the *Science of Charms* — written between 2000 and 1400 B.C. — as one of the, "five kingdoms of herbs . . . which release us from anxiety." Bhang seems to have been popular with the Indian people from the beginning of history. The Indian Hemp Drugs Commission Report

(1894) gave a detailed picture of how bhang and the more potent cannabis products ganja and *charas* (the Indian term for cannabis resin) had become incorporated into Indian life and culture.

It took longer for cannabis to reach the West. Hemp was known to the Assyrian civilization both as a fiber plant and a medicine and is referred to as *kunnubu* or *kunnapu* in Assyrian documents of around 600 B.C. The word is probably the basis of the Arabian *kinnab* and the Greek and Latin *cannabis*. There is little evidence that the plant was known beyond Turkey until the time of the Greeks. The Greeks used hemp for the manufacture of ropes and sails for their conquering navies, as did the Romans later—although the hemp was not cultivated in Greece or Italy but in the further reaches of their empires in Asia Minor. Neither the Greeks nor the Romans, however, appear to have used cannabis for its psychoactive properties, although these were known and described by the Roman physicians Dioscorides, Galen and Oribasius. Galen writing in the second century A.D. described how wealthy Romans sometimes offered their dinner guests an exotic dessert containing cannabis seeds:

> There are those who eat it (cannabis seed) also cooked with other confections, by this confection is meant a sort of dessert which is taken after meals with drinks for the purpose of exciting pleasure. It creates much warmth (or possibly excitement) and when taken too generously affects the head emitting a warm vapor and acting as a drug.
>
> (Walton, 1938, p.8)

As the seeds contain no significant amounts of psychoactive material it seems likely that some other parts of the cannabis plant must also have been included.

It was to be almost another thousand years before cannabis spread to the Arab lands and then to Europe and the Americas. According to one Arab legend, the discovery of marijuana dates back to the twelfth century A.D. when a monk and recluse named Hayder, a Persian founder of the religious order of Sufi, came across the plant while wandering in meditation in the mountains. When he returned to his monastery after eating some cannabis leaves, his disciples were amazed at how talkative and animated this normally dour and taciturn man had become. After they persuaded Hayder to tell them what had made him so happy, the disciples went out into the mountains and tried some cannabis themselves. By the thirteenth century cannabis use had become common in the Arab lands, giving rise to many colorful legends. The *Arabian Nights* and *The Thousand and One Nights* are folk tales collected during the period 1000–1700 A.D. Bhang (marijuana) and hashish are referred to frequently:

> Furthermore, I conceive that the twain are eaters of Hashish, which drug when swallowed by man, maketh him prattle of whatso he pleaseth and chooseth, making him now a Sultan, then a Wazir, and then a merchant, the while it seemeth to him that the world is in the hollow of his hand. Tis composed of hemp leaflets whereto are added aromatic roots and somewhat of sugar; then they cook it and prepare a kind of confection which they eat, but whoso eateth it (especially if he eat more than enough) talketh of matters which reason may on no wise represent.
>
> (Walton, 1938, p.15)

It is clear from this description that the word hashish in ancient Arab writings refers to what we would now call marijuana rather than the cannabis resin that the term hashish now describes. Outstanding among the Arab legends is the story of the *Old Man of the Mountains* and his murderous band of followers known as the *Assassins*. According to Marco Polo who recorded this legend, the Assassins were lead by the Old Man of the Mountains who recruited novices to his band and kept them under his control as his docile servants by feeding them copious amounts of hashish. Marco Polo described how the leader constructed a remarkable garden at his major fortress, the Alamut. The young assassins would be transported to the garden after they had taken enough hashish to put them to sleep. When they awoke, and found themselves in such a beautiful place with ladies willing to dally with them to their heart's content, they believed that they were indeed in paradise. When the Old Man wanted someone killed he would tell the assassins to do it, and promise them that dead or alive they would return to paradise.

Although the historical facts are impossible to determine, it seems likely that the Assassins were lead by Hasan-Ibn-Sabbah, who started life as a religious missionary and later gathered a secret band of followers. They probably used hashish, as did many others in the Arab world at that time. It seems less likely that they would have been able to carry out their terrorist acts while intoxicated by cannabis, nor is there any significant evidence that the drug inspires violence — on the contrary it tends to cause somnolence and lethargy when taken in high doses. Nevertheless, lurid stories about the drug-crazed Assassins have been used widely in the West as part of the mythology that surrounds the cannabis debate. As early as the twelfth century Abbot Arnold of Lübeck wrote in *Chronica Slavorum*:

> Hemp raises them to a state of ecstasy or folly, or intoxicates them. Then sorcerers draw near and exhibit to the sleepers phantasms, pleasures and amusements. They then promise that these delights will become perpetual if the orders given them are executed with the daggers provided.

Eight hundred years later in the United States the hard line commissioner of the Federal Bureau of Narcotics, Harry J. Anslinger, used the image of the drug-crazed assassin in his personal vendetta against the drug. He wrote in the *American Magazine* in 1937:

> In the year 1090, there was founded in Persia the religious and military order of the Assassins, whose history is one of cruelty, barbarity and murder, and for good reason. The members were confirmed users of hashish, or marijuana, and it is from the Arab "hashishin" that we have the English word "assassin."

> (Anslinger and Cooper, 1937)

The use of cannabis was particularly common in Egypt in the Middle Ages where the Gardens of Cafour in Cairo became a notorious haunt of hashish smokers. Despite draconian measures by the Egyptian authorities to close such establishments and to prohibit hashish use during the thirteenth and fourteenth centuries the habit had become too firmly ingrained in the Arab world for it to be stamped out. The social acceptance of cannabis use among the people of Egypt and other Arab lands was reinforced by the fact that while the holy Koran explicitly banned the consumption of alcohol, it did not mention cannabis. Not all were happy about this acceptance of cannabis, however. Ebn-Beitar wrote of the spread of cannabis use in Egypt 600 years ago:

> It spread insensibly for several years and became of common enough usage that in the year 1413 A.D., this wretched drug appeared publicly, it was eaten flagrantly and without furtiveness, it triumphed. . . . One had no shame in speaking of it openly. . . . Also as a consequence of that, baseness of sentiments and manners became general; shame and modesty disappeared among men, they no longer blushed to hold discourse on the most indecent things. . . . And they came to the point of glorifying vices. All sentiments of nobility and virtue were lost. . . . And all manner of vices and base inclination were displayed openly.
>
> (Walton, 1938, p.14)

It was from Egypt that the use of cannabis as a psychoactive drug first spread to Europe and then to the Americas. When Napoleon invaded and conquered Egypt at the end of the eighteenth century he was dismayed by what he saw as the corrupting influence of hashish on the local population and the possible debilitating effects it might have on his own soldiers. One of his generals issued a decree:

> Article 1: Throughout Egypt the use of a beverage prepared by some Moslems from hemp (hashish) as well as the smoking of the seeds of hemp, is prohibited. Habitual smokers and drinkers of this plant lose their reason and suffer from violent delirium in which they are liable to commit excesses of all kinds.
>
> Article 2: The preparation of hashish as a beverage is prohibited throughout Egypt. The doors of those cafes and restaurants where it is supplied are to be walled up, and their proprietors imprisoned for three months.
>
> Article 3: All bales of hashish arriving at the customs shall be confiscated and burnt.
>
> (Lewin, 1931)

As with all the earlier bans this one too was largely ignored by the Egyptians and Napoleon's army was soon to leave in retreat. However, the returning French army brought back to Europe many colorful tales of hashish and its intoxicating effects. Although cannabis had been cultivated in Europe for many centuries as a source of rope, canvas, and other

cloths and in making paper, its inebriating effects were largely unknown — although secretly some sorcerers and witches may have included cannabis in their mysterious concoctions of drugs. In the midnineteenth century in France it became fashionable among a group of
writers, poets, and artists in Paris's latin quarter to experiment with hashish. Among these was the young French author Pierre Gautier who became so enthused by the drug that he founded the famous Club des
Hashischins in Paris and introduced many others among the French literary world to its use. These included Alexander Dumas, Gerard de Ner-
val, and Victor Hugo — all of whom wrote about their experiences with
hashish. Gautier and his sophisticated literary colleagues regarded cannabis as an escape from a bourgeois environment, and described their
drug-induced experiences in flowery, romantic language. Thus, Gautier
wrote the following:

> After several minutes a sense of numbness overwhelmed me. It seemed that
> my body had dissolved and become transparent. I saw very clearly inside
> me the Hashish I had eaten, in the form of an emerald which radiated
> millions of tiny sparks. The lashes of my eyes elongated themselves to In-
> finity, rolling like threads of gold on little ivory wheels, which spun about
> with an amazing rapidity. All around me I heard the shattering and crum-
> bling of jewels of all colours, songs renewed themselves without ceasing, as
> in the play of a kaleidoscope.
>
> (Walton, 1938, p.59)

Among the most influential of Gautier's colleagues was Charles
Baudelaire, whose book *Les Paradis Artificiels* published in Paris in 1860
described the hashish experience in romantic and imaginative language:

> . . . the senses become extraordinarily acute and fine. The eyes pierce
> Infinity. The ear perceives the most imperceptible in the midst of the
> sharpest noises. Hallucinations begin. External objects take on monstrous
> appearances and reveal themselves under forms hitherto unknown. They
> then become deformed and at last they enter into your being or rather
> you enter in to theirs. The most singular equivocations, the most inexpli-
> cable transpositions of ideas take place. Sounds have odour and colours
> are musical.

The book captured the imagination of many readers in the West and inspired further interest in the use of cannabis; it is still one of the most comprehensive and impressive accounts of the effects of cannabis on the human psyche. The use of hashish, however, did not become widespread in Europe. Cannabis use was practically unknown in Britain, for example, until the 1960s, although hemp had been cultivated for hundreds of years as a fiber and food crop. Similarly, in North America the hemp plant was imported shortly after the first settlements and was widely cultivated. Kentucky was particularly renowned for its hemp fields, and "Kentucky Hemp," selected for its fiber production, is an important fiber variety. Americans seemed unaware of the peculiar properties of cannabis, and it is also unlikely that the varieties selected for hemp fiber production contained significant amounts of THC. It was not until the well-known midnineteenth century American author Bayard Taylor wrote a lurid account of his experiences with hashish in the Middle East that there was any awareness of the psychoactive effects of cannabis. Taylor described what happened after taking a large dose of the drug:

> The spirit (demon, shall I not rather say?) of Hasheesh had entire possession of me. I was cast upon the flood of his illusions, and drifted helplessly withersoever they might choose to bear me. The thrills which ran through my nervous system became more rapid and fierce, accompanied with sensations that steeped my whole being in inutterable rapture. I was encompassed in a seal of light, through which played the pure, harmonious colours that are born of light . . . I inhaled the most delicious perfumes; and harmonies such as Beethoven may have heard in dreams but never wrote, floated around me.
>
> (Walton, 1938, p.65)

Taylor's accounts were intentionally sensational and played to the nineteenth century appetite for tales of adventure and vice in far away places. It is unlikely that many readers were encouraged to experiment with cannabis themselves. One exception, however, was a young man named Fitz Hugh Ludlow. Ludlow experimented with many drugs, and started taking cannabis, then widely available in the United States in various pharmaceutical preparations. Ludlow's detailed accounts of his experiences, and his subsequent addiction to cannabis are described in detail in his book *The Hasheesh Eater*. Ludlow was an intelligent youth of 16 when he discovered cannabis in the local drug store where he had already experimented with ether, chloroform, and opium. He used cannabis intensely for the next 3 or 4 years, and wrote of his experiences as part of his withdrawal from the drug. The book has become a classic in the cannabis literature, equivalent in importance to Baudelaire's *Les Paradis Artificiels*, and is referred to again in Chapter 3. Ludlow's book, however, seems to have had little impact at the time of its publication. One reviewer of his book, writing in 1857 commented that America was fortunately, "in no danger of becoming a nation of hasheesh eaters."

For almost a hundred years from the midnineteenth century until 1937 cannabis enjoyed a brief vogue in Western medicine (see Chapter 4). Following its introduction from Indian folk medicine, first to Britain and then to the rest of Europe and to the United States a variety of different medicinal cannabis products were used.

The cannabis plant was introduced to Latin America and the Caribbean as early as the first half of the fifteenth century by slaves brought from Africa. It became fairly widely used in many countries in this region for its psychoactive properties, both as a recreational drug and in connection with various native Indian religious rites (see Chapter 6). The term "marijuana," a Spanish–Mexican word originally used to describe tobacco, came into general use to describe cannabis in both South and North America.

The history of marijuana use in the United States and its prohibition has been told many times (Snyder, 1971; Abel, 1943). After a brief vogue in the midnineteenth century, the popularity of marijuana waned, and it was only regularly used in the United States in a few large cities by local groups of Mexicans and by African-American jazz musicians. It was the wave of immigrants who entered the southern United States from Mexico in the early decades of the twentieth century, bringing marijuana with them, that first brought the drug into prominence in America — and lead to its prohibition. It came initially to New Orleans and some other southern cities and spread slowly in some of the major cities. There were colorful accusations that marijuana use provoked violent crime and corrupted the young. The head of the Federal Narcotics Bureau, Harry Anslinger waged an impassioned campaign to outlaw the drug. He was the original spin doctor of his time, cleverly manipulating other government agencies, popular opinion, and the media with lurid tales of the supposed evils of cannabis. In 1937, the United States Congress, almost by default, passed the Marijuana Tax Act, which effectively banned any further use of the drug in medicine and outlawed it as a dangerous narcotic. Use of the drug continued to grow, however, and by the late 1930s newspapers in many large cities were filled with alarming stories about this new "killer drug."

In 1937, no less than 28 different pharmaceutical preparations were available to American physicians, ranging from pills, tablets, and syrups containing cannabis extracts, to mixtures of cannabis with other drugs — including morphine, chloroform, and chloral. American pharmaceutical companies had begun to take an active interest in research on cannabis-based medicines. The hastily approved Cannabis Tax Act put a stop to all further medical use and essentially terminated all research in the field for another 25–30 years.

The "demonization" of cannabis in the United States soon after its arrival from Latin America has colored attitudes to the drug ever since — not only in North America but world wide. In subsequent chapters the reader can judge whether this initial reaction to cannabis was justified.

2

*The Pharmacology of THC,
the Psychoactive Ingredient
in Cannabis*

As cannabis came into widespread use in Western medicine in the nineteenth century it soon became apparent that the effects of plant-derived preparations were erratic. The amounts of active material that the pharmaceutical preparations contained were variable from batch to batch according to the origin of the material, the cultivation conditions, and the plant variety. As the chemical identity of the active ingredients was not known and there was no method of measuring them, there was no possibility of quality control. This was one of the reasons cannabis preparations eventually fell out of favor with physicians on both sides of the Atlantic. These inadequacies, however, also motivated an active research effort to identify the active principles present in the plant preparations in the hope that the pure compound or compounds might provide more reliable medicines. The nineteenth century was a great era for plant chemistry. Many complex drug molecules, known as alkaloids, were isolated and identified from plants. Several of these were powerful poisons, e.g., atropine from deadly nightshade (*Atropa belladonna*), strychnine from the bark of the tree nux vomica, and muscarine from the magic mushroom, *Amanita muscaria*. Others were valuable medicines, still in use today, e.g., morphine isolated from the opium poppy, *Papaver somniferum*; the antimalarial drug quinine from the bark of the South American cinchona tree; and cocaine from the leaves of the coca plant. Victorian chemists were attracted by the new challenge offered by isolating the active ingredient from cannabis and attacked the problem with vigor, but initially without any notable success. Unlike the previously discovered alkaloids, which were all water soluble organic bases that could form crystalline solids when combined with acids, the active principle of the cannabis plant proved to be almost completely insoluble in water. The active compound, tetrahydrocannabinol (THC), is a viscous resin with no acidic or basic properties, so it cannot be crystallized. Since most of the previous successes of natural product chemistry had depended on the ability of chemists to extract an active drug substance from the plant with acids or alkalis and to obtain it in a pure crystalline form, it was not surprising that all of the early efforts to find the cannabis alkaloid in this way were doomed to failure. Only those who recognized that the active principle could not be extracted into aqueous solutions but required an organic solvent (usually alcohol) were able to make any real progress. T. & H. Smith, brothers who founded a pharmaceutical business based on medicinal plant extracts in Edinburgh in the midnineteenth century described in 1846 how they extracted Indian ganja repeatedly with warm water and sodium carbonate alkali to remove the water soluble plant materials, and then extracted the remaining dried ganja residue with absolute alcohol. The alcoholic extract was treated successively with alkaline milk of lime and with sulphuric acid and then evaporated to leave a small amount of viscous resin (6%–7% of the weight of the starting material) to which they gave the name *cannabin*. It was clear from the nature of the procedures used that the resin was neither acid nor base but neutral. The purified resin proved to be highly active when tested in the then traditional manner on themselves:

> two thirds of a grain (44 mg) of this resin acts upon ourselves as a powerful narcotic, and one grain produces complete intoxication.
>
> (Smith and Smith, 1846)

The British chemists Wood, Spivey, and Easterfield working in Cambridge, England at the end of the nineteenth century made another important advance (see review by Todd, 1946). They used Indian charas (cannabis resin) as their starting material and extracted this with a mixture of alcohol and petroleum ether. From this by using the then new technique of fractional distillation they isolated a variety of different materials, including a red oil or resin of high boiling point (265°C) that was toxic in animals and that they suspected to be the active ingredient; they named it *cannabinol*. A sample of the purified material was passed to the professor of medicine in Cambridge for further pharmacological investigation. The report published in *Lancet* in 1897 by his research assistant Dr. C. R. Marshall (Marshall, 1897) illustrates the heroic nature of pharmacological research in that era. He described his experience on taking a sample of the material as follows:

> On the afternoon of Feb 19th last, whilst engaged in putting up an apparatus for the distillation of zinc ethyl, I took from 0.1 to 0.15 gramme of the pure substance from the end of a glass rod. It was about 2.30 P.M. The substance very gradually dissolved in my mouth; it possessed a peculiar

pungent, aromatic, and slightly bitter taste, and seemed after some time to produce a slight anaesthesia of the mucous membranes covering the tongue and fauces. I forgot all about it and went on with my work. Soon after the zinc ethyl had commenced to distil — about 3.15 — I suddenly felt a peculiar dryness in the mouth, apparently due to an increased viscidity of the saliva. This was quickly followed by paraesthesia and weakness in the legs, and this in turn by diminution in mental power and a tendency to wander aimlessly about the room. I now became unable to fix my attention on anything and I had the most irresistible desire to laugh. Everything seemed so ridiculously funny; even circumstances of a serious nature were productive of mirth. When told that a connection was broken and that air* was getting into the apparatus and an explosion feared I sat upon the stool and laughed incessantly for several minutes. Even now I remember how my cheeks ached. Shortly afterwards I managed to collect myself suffi-ciently to aid in the experiment, but I soon lapsed again into my former state. This alternating sobriety and risibility occurred again and again, but the lucid intervals gradually grew shorter and I soon fell under the full influence of the drug. I was now in a condition of acute intoxication, my speech was slurring, and my gait ataxic. I was free from all sense of care and worry and consequently felt extremely happy. When reclining in a chair I was happy beyond description, and afterwards I was told that I constantly exclaimed, "This is lovely!" But I do not remember having any hallucinations: the happiness seemed rather to result from an absence of all external irritation. Fits of laughter still occurred; the muscles of my face being sometimes drawn to an almost painful degree. The most peculiar effect was a complete loss of time relation: time seemed to have no exis-tence: I appeared to be living in a present without a future or a past. I was constantly taking out my watch thinking hours must have passed and only a few minutes had elapsed. This, I believe, was due to a complete loss of memory for recent events. Thus, if I walked out of the room I should return immediately, having completely forgotten that I had been there be-fore. If I closed my eyes I forgot my surroundings and on one occasion I asked a friend standing near how he was several times within a minute. Between times I had merely closed my eyes and forgotten his existence.

*Zinc ethyl burns on contact with air and consequently must be distilled in an atmosphere of carbon dioxide.

(Marshall, 1897)

Marshall's colleagues became increasingly worried about him and eventually sent for medical help, but by the time the doctor arrived at around 5:00 P.M. Marshall had begun to recover and by 6:00 P.M. he was on his way home after a cup of coffee and suffered no ill effects afterwards. Despite his experience, Marshall volunteered to take another dose of the resin 3 weeks later, but this time a much smaller one (50 mg). This produced essentially the same symptoms but in a milder form. It is clear that the red oil isolated by Wood and colleagues was highly enriched in the active component or components of cannabis, and Marshall's description accurately describes the typical intoxication seen after high doses of the drug.

Although Wood and his colleagues in Cambridge had come close to purifying the active ingredient in cannabis their further work turned out to lead them down a blind alley. From the red oil they were able to isolate a crystalline material after the preparation was acetylated (which produced acetyl derivatives of any compound with a free hydroxyl (-OH) group). After purification of this crystalline derivative and removal of the acetyl groups by hydrolysis they succeeded in isolating a compound that they called *cannabinol* and they showed that it could apparently be extracted from various other cannabis products, including several of the cannabis-containing medicines then available. The earlier red oil fraction was now renamed *crude cannabinol*. Unfortunately, however, cannabinol was not the active ingredient but a chemical degradation product formed either during the chemical purification procedures, or present as a normal degradation product in samples of cannabis material that had been stored for too long. The findings made with the original red oil material must have been due to the presence of THC in such samples. It was believed, erroneously for decades after this that cannabinol was indeed the active principle of cannabis, although other laboratories were unable to repeat the findings of Wood and his colleagues.

Thirty years later a brilliant young British chemist Cahn revisited the problem of cannabinol (see review by Todd, 1946). He was able to isolate the pure substance as described by Wood and colleagues, and using the improved chemical techniques available in the 1920s he carried out a meticulous series of experiments that largely established the

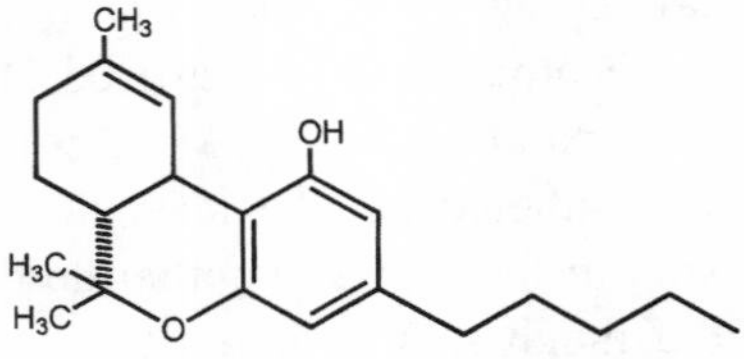

delta-9-THC

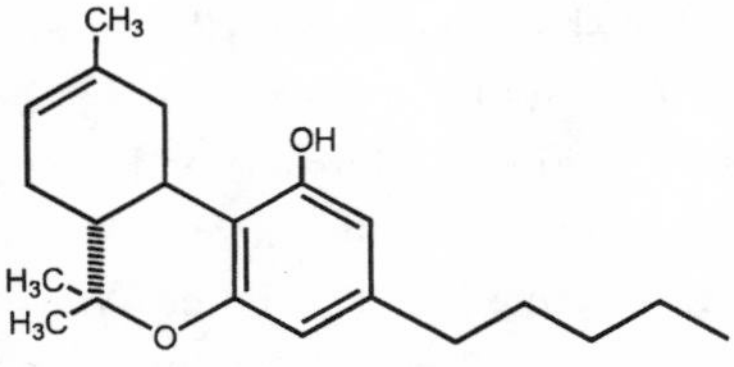

delta-8-THC

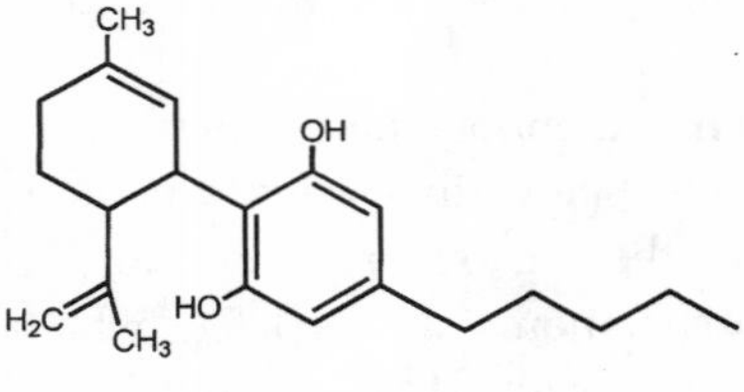

Cannabidiol

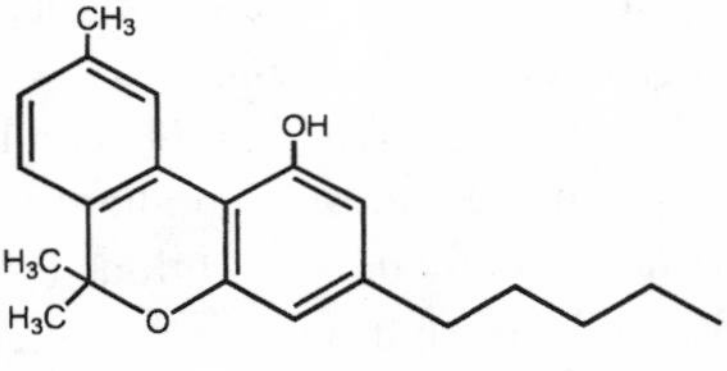

Cannabinol

Figure 2.1. Naturally occurring cannabinoids in cannabis extracts; delta-9-THC is the main psychoactive ingredient.

chemical structure of cannabinol (Fig. 2.1). Although this was not the true active principle, the new structure allowed chemists to synthesize a range of related compounds and Cahn's work provided a great impetus to further chemistry research in this field.

At the University of Illinois in the 1940s Roger Adams was also working on the problem (Adams, 1942). He used an alcoholic extract from which he produced a red oil by distillation. From this he was able to purify a crystalline benzoic acid derivative of a compound that he named *cannabidiol* (as it contained two hydroxyl groups), and to work out its chemical structure (Fig. 2.1). This was a real advance, as this compound — unlike the cannabinol worked on by Wood and colleagues — really is one of the naturally occurring materials in the cannabis plant. Unfortunately though it is not the active ingredient and the narcotic activity that was reported by volunteers who took samples of

Adams's cannabidiol must have been due to contamination with THC. Nevertheless, Adams and his group were able to synthesize various chemical derivatives of cannabidiol, including hydrogenated derivatives, the tetrahydrocannabinols, and some of these did possess potent psychoactive properties (measured both in human volunteers and increasingly by observing the behavioral responses of rodents, dogs, and other laboratory animals). In his 1942 *Harvey Lecture* Adams wrote:

> The typical marijuana activity manifested by the isomeric tetrahydrocannabinols constitutes ponderable evidence that the activity of the plant itself, and of extracts prepared therefrom, is due in large part to one or other of these compounds . . .

(Adams, 1942)

At the same time, across the Atlantic, despite the privations of war, research on cannabinoids continued in the chemistry department in Cambridge England under the leadership of an outstanding organic chemist Alexander Todd, later to become Lord Todd. He and his colleagues reisolated cannabinol, and capitalizing on the newly discovered structure of cannabidiol published by the Adams group, they were able to complete the identification of the chemical structure of this compound started by Cahn (Todd, 1946). Both the Adams group and the Todd group went on to undertake the first chemical synthesis of cannabinol, and as part of this synthesis the Cambridge team actually made delta-9-tetrahydrocannabinol (THC) as an intermediate from which to form cannabinol. They commented on the high degree of biological activity that this compound possessed (assessed now by observing the characteristic behavioral reactions of dogs and rabbits rather than human subjects). The Todd group repeatedly tried to prove that this compound or something like it existed naturally in cannabis extracts. By repeated fractionation they were able to prepare a highly active and almost colorless glassy resin that closely resembled synthetic tetrahydrocannabinol in its physical and chemical properties. The techniques available then, however, were not powerful enough to determine whether this was a single chemical substance or a complex mixture of closely related compounds. In a review article published in 1946 Todd wrote:

. . . it would appear to be established that the activity of hemp resin, in rabbits and dogs at least, is to be attributed in the main to tetrahydro-cannabinols.

(Todd 1946)

Tetrahydrocannabinol was also isolated from a red oil fraction by the American chemist Wollner in 1942 though not as a single pure compound but as a mixture containing tetrahydrocannabinols. It was assumed for many years after the advances of the 1940s that the psychoactive properties of cannabis were due to an ill-defined mixture of such compounds. It was to be another 20 years before the brilliant chemical detective work of two Israeli scientists, Mechoulam and Gaoni finally solved the problem and showed that in fact there is only one major active component, delta-9-tetrahydrocannabinol (THC) (Fig. 2.1; Mechoulam, 1970). Raphael Mechoulam described their introduction to this field as follows:

When we started our then very small programme on hashish some 5–6 years ago, our interest in this fascinating field was kindled by the contrast of rich folklore and popular belief with paucity of scientific knowledge. Israel is situated in a part of the world where, for many, hashish is a way of life. Though neither a producer nor a large consumer, Israel is a crossroads for smugglers, mostly Arab Bedouin, who get Lebanese hashish from Jordan through the Negev and Sinai deserts to Egypt. Hence the police vaults are full of material waiting for a chemist.

(Mechoulam et al., 1970)

Gaoni and Mechoulam had the advantage of new chemical separation and analytical techniques that had not been available to earlier investigators. In the laboratory of natural products at the Hebrew University in Jerusalem they had the latest methods for separating complex mixtures of chemicals by column chromatography. In this technique the mixture is poured onto a column of adsorbent material and gradually washed through by solvents. Individual compounds move down the column at different rates according to how easily they dissolve in the solvent flowing through the column. In addition the Israeli scientists were able to employ the powerful new techniques of mass spectrometry, infrared spectroscopy and nuclear magnetic resonance to identify the chemicals that they had separated by chromatography. In this way they were able to identify a large number of new cannabinoids in extracts of Lebanese hashish — we now know that as many as 60 different naturally occurring cannabinoids exist. Although this complexity might appear daunting, it turned out that most of the naturally occurring cannabinoids were present in relatively small amounts, or that they lacked biological activity. In fact Gaoni and Mechoulam reported in 1964 that virtually all of the pharmacological activity in hashish extracts could be attributed to a single compound delta-9–tetrahydrocannabinol (THC).[1]

Among other chemicals in the hashish extracts Gaoni and Mechaloum identified cannabidiol (Fig. 2.1). They found a variety of other naturally occurring cannabinoids, but delta-9-THC was the most important. Cannabidiol is present in significant quantities but lacks psychoactive properties, although it may have other pharmacological effects (see Chapter 2). Cannabis grown in tropical parts of the world (Africa, Southeast Asia, Brazil, Colombia, Mexico) usually has much more THC than cannabidiol, with ratios of THC/cannabidiol of 10:1 or higher. Plants grown outdoors in more northern latitudes, however, (Europe, Canada, northern United States) usually have a much higher content of cannabidiol, often exceeding the THC content by 2:1 (Clarke, 1981, p.159). Cannabis also contains variable amounts of carboxylic acid derivatives of delta-9-THC, and this is potentially important. Although themselves inactive, the carboxylic acid derivatives readily lose their carboxylate group as carbon dioxide on heating to form active THC. This occurs, for example, when the plant material is heated during smoking, or heated in the cooking processes used to form various cannabis-containing foods and drinks. This can in some instances more than double the active THC content of the original starting plant material. On the other hand, when cannabis resin or other preparations are stored, pharmacological activity

---

1. In some publications, including those from the Israeli group, this is referred to as delta-1-tetrahydrocannabinol, but this is because there are two different conventions for numbering the chemical ring systems of which the substances are composed; the delta-9 terminology is the most commonly used.

38 THE SCIENCE OF MARIJUANA

is gradually lost and THC degrades by oxidation to cannabinol and other inactive materials.

The isolation and elucidation of the structure of delta-9-THC led to a burst of chemical synthetic activity around the world, as different laboratories competed to be the first to complete the synthesis of this important new natural product. The American chemists Taylor, Lenard, and Shvo were probably the first in 1967, but they were quickly followed by Gaoni and Mechoulam and by several other laboratories (for review see Mechoulam et al., 1970). The Israeli group had shown that the naturally occurring THC occurred only as the *l*-isomer, although early synthetic preparations contained a mixture of both the *l*- and *d*-optical isomers (mirror images) of the compound. So the next stage was for several laboratories to devise chemical synthetic methods that yielded only the naturally occurring *l*-isomer of delta-9-THC, which is biologically far more active than the mirror image *d*-isomer.

In retrospect, although the isolation of THC from cannabis proved technically difficult because of the nature of the compound as a neutral, water insoluble, viscous resin, the outcome was not very different from that seen with the isolation of other pharmacologically active substances from plants. In each case a single active compound has been identified that represents virtually all of the biological activity in the crude plant extracts, although the active material often exists in the plant as one member of a complex mixture of related chemicals, most of which are either minor components or lack biological activity. This is true, for example, for nicotine from the tobacco leaf, cocaine from the coca leaf, and morphine from the opium poppy.

Man-Made Cannabinoids

The synthesis of THC was followed by a much larger synthetic chemistry effort, aimed at the discovery of more potent analogues of THC, or compounds that separated the desirable medical properties of THC from its psychoactive effects. Many hundreds of new THC derivatives were made during the 1950s and 1960s in both academic and pharmaceutical company laboratories. There were far too many to be tested on human volunteers, so most were assessed in simple animal behavior tests that had been found to predict cannabis-like activity in man (see Chapter 3). This research effort was disappointing because it proved impossible to separate the desirable properties of THC (antinausea, pain relieving) from the intoxicating effects. Nevertheless, the chemical research provided a detailed insight into the structure activity of the THC molecule, i.e., which parts of the molecule are critical for psychoactivity, and which parts are less important and can thus be chemically modified without losing biological activity. Several derivatives proved to be even more active than THC, working in animals and human volunteers at doses up to 100 times lower than required for THC (for review see Duane Sofia, 1978)

At the Pfizer company in the United States, for example, chemists were among the first to discover the first potent synthetic THC analogue *nantradol*, which entered pilot scale clinical trials and was found to have analgesic (pain-relieving) properties that were not blocked by the drug naloxone — an antagonist that blocks analgesics of the morphine type that act on opiate receptors. Nantradol as synthesized originally was a mixture of four chemical isomers from which the active one *levonantradol* was later isolated. These compounds had an important advantage over THC in being water soluble and thus easier to formulate and to deliver. Further chemical work at Pfizer lead to the discovery of a new chemical series of simplified THC analogues that possessed only two of the three rings of THC, among these bicyclic compounds was the potent analogue CP-55,940 (Fig. 2.2), which has been widely used as a valuable research compound. The Pfizer compound levonantradol was tested in several clinical trials during the early 1980s. It proved to be considerably more potent than morphine as an analgesic, and was effective in blocking nausea and vomiting in patients undergoing cancer chemotherapy. Nevertheless, the psychoactive side effects proved to be unacceptable and the company decided to abandon further research on this project (Dr. Ken Coe, personal communication).

Work in Raphael Mechoulam's laboratory in Israel was particularly productive in generating new analogues of THC (e.g., HU-210, Fig. 2.2). And research in the pharmaceutical company Eli Lilly in the United

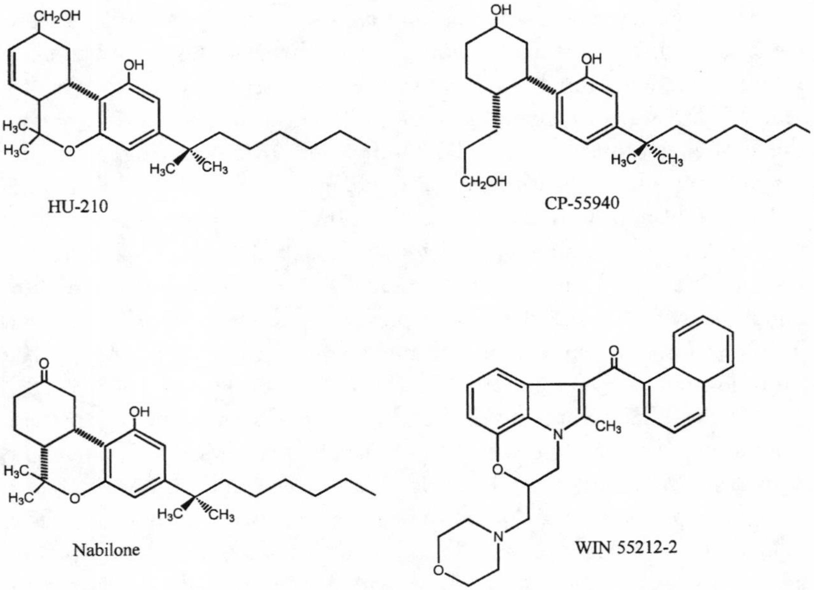

Figure 2.2. Man-made synthetic cannabinoids.

States at this time lead to the synthesis of *nabilone* (Fig. 2.2), the only synthetic THC analogue that has been developed and approved as a medicine (Chapter 4).

In a surprising development, research scientists at the Sterling Drug Company in the United States unwittingly discovered another chemical class of molecules that did not immediately resemble THC, but nevertheless proved to act through the same biological mechanisms. A research program aimed at discovering novel aspirin-like antiinflammatory/ pain relieving compounds generated an unusual lead compound called *pravadoline*. This had a remarkable profile in animal tests — it was highly effective in a broad range of pain tests — including ones in which aspirin-like molecules generally do not work. In addition it failed to cause any gastric irritation, one of the biggest drawbacks in the aspirin class of drugs. Nor was pravadoline very effective in the key biochemical test for aspirin-like activity, the ability to inhibit the synthesis of the inflammatory chemicals the prostaglandins. It seemed to the scientists involved that they had discovered a promising new mechanism for pain relief— and one that might have important advantages. Pravadoline went into clinical development, and meanwhile many other analogues were synthesized. From these emerged the compound WIN 55212-2 (Fig. 2.2; D'Ambra et al., 1996), an even more potent pain-relieving compound with improved absorption properties. However, when the specific receptor for cannabis was discovered in the 1980s (see below) it became clear that pravadoline and WIN 55212-2 acted like THC on this receptor (Kuster et al., 1993), and were thus in pharmacological terms cannabinoids rather than aspirin-like antiinflammatory drugs. Their pain relieving properties were not due to a new mechanism but to the same mechanism as that of cannabis. Pravadoline had by that time been tested in human volunteers, and found to possess good effectiveness against moderate to severe pain in, for example, postoperative dental pain. But it also caused dizziness and light-headedness as an obvious limiting side effect. The development of pravadoline was dropped because of kidney toxicity, and the company then decided to abandon the whole program—partly for budget reasons and partly to avoid being associated with the image of a cannabis-like drug (Dr. Susan Ward, personal communication).

Structure–Activity Relationships for Cannabinoids

Since the elucidation of the chemical structure of THC in 1964 many hundreds of chemical analogues have been synthesized and tested. In the period before the discovery of the cannabinoid CB-1 receptor and the development of simple test tube receptor binding and adenylate cyclase assays these analogues had to be assessed in human volunteers or in whole animal experiments. Many of the laboratories involved relied heavily on the use of such large animals as the rhesus monkey or dog, and several laboratories later adopted the simple group of behavioral tests devised by Martin and his colleagues using mice (Abood and Martin, 1992) (see Fig. 2.12). Remarkably, despite the use of this variety of phar-

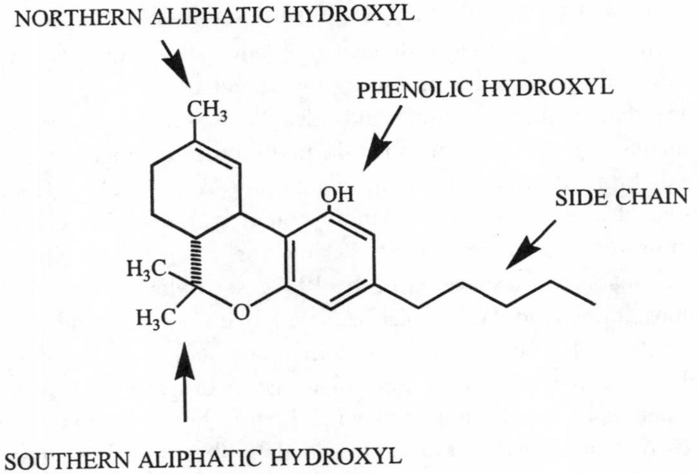

NORTHERN ALIPHATIC HYDROXYL

PHENOLIC HYDROXYL

SIDE CHAIN

SOUTHERN ALIPHATIC HYDROXYL

Figure 2.3. The key structural elements that contribute to biological activity in THC and synthetic cannabinoids.

macological models, a consistent body of evidence was built up that defined the chemical structure activity rules that determine whether a molecule will be active at the CB-1 receptor (for reviews see Mechoulam et al., 1970; Duane Sofia, 1978; Makriyannis and Rapaka, 1990).

At least four molecular fragments of the tetrahydrocannabinol structure contribute to cannabimimetic activity (Fig. 2.3). The phenolic hydroxyl on the A ring is necessary for cannabinoid activity. Elimination, substitution with an alkyl group, esterification of the hydroxyl, ether derivatives, or replacement of the oxygen with another heteroatom eliminates all activity. The side chain is crucial. Potency can be increased up to a point by an increase in chain length with a 7-carbon chain being optimal. Methyl substitution at the side-chain carbon adjacent to the aromatic ring also enhances activity. Analogues with substituents in the A ring that are ortho to the phenolic hydroxyl group retain substantial activity while substituents that are para to the hydroxyl group lose all activity. This suggests that the orientation of the side chain in a southern direction plays an important role.

The presence of a northern aliphatic hydroxyl can result in enhanced activity, as in the major metabolite of THC, 11-hydroxy-THC in which the 11-methyl group is oxidized to a hydroxymethyl. The configuration of this hydroxyl group is critical in determining potency, with the equatorial β-isomer being far more potent than the axial α-isomer. This rule also applies to the nonclassical synthetic cannabinoids. These lack a tetrahydropyran ring and possess a longer 1,1-dimethylheptyl side chain; the most potent compounds have a southern aliphatic hydroxyl, sometimes attached to a cyclohexyl component that replaces the C-ring of THC. In the C-ring double bond isomers the order of potency is delta-9-THC = delta-8-THC > delta-10-THC. Delta-7-THC and delta-11-THC are inactive.

Much less medicinal chemistry research has been done so far on the requirements for activity at the CB-2 receptor, but it is already clear that selective agonists and antagonists can be developed that have relatively little overlap in their activity at the CB-1 receptor. For example, the phenolic hydroxyl on the A-ring can be methylated without loss of CB-2 activity.

Cannabinoid Antagonists

An important recent development has been the discovery of molecules that bind to the cannabis receptor in the brain but instead of mimicking THC they block its actions. Like the synthetic cannabinoids these come from various different chemical classes and the three CB1 receptor antagonists currently known are shown in Figure 2.3. The first cannabinoid antagonist to be described was the compound SR141716A from the French pharmaceutical company Sanofi and this has been used extensively in the past few years as a research compound. Other compounds with CB1 antagonist activity have since been described by other pharmaceutical companies (Fig. 2.4). Subsequently the Sanofi compound SR144528 became available as the first selective antagonist acting at CB-2 receptors. The antagonists, as we shall see, represent valuable new research tools with which to ask questions about the normal functions of

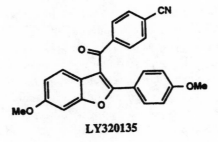

LY320135

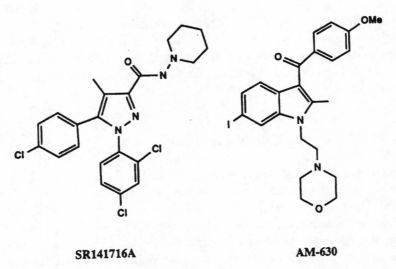

SR141716A   **AM-630**

Figure 2.4. Synthetic drugs that act as antagonists at the CB_1 cannabinoid receptor.

the cannabinoid systems in the body and the extent to which long-term use of cannabis may lead to the development of physical dependence on the drug.

How Does THC Get to the Brain?

Smoking

Smoking is an especially effective way of delivering psychoactive drugs to the brain. When marijuana is smoked some of the THC in the burning plant material distils into a vapor (the boiling point of the THC resin is 200° C) and as the vapor cools the THC condenses again into fine droplets, forming a smoke that is inhaled. As the drug dissolves readily in fats, it passes quickly through the membranes lining the lungs, which offer a large surface area for absorption. The drug enters the blood, which passes directly from the lungs to the heart from where it is pumped in the arteries around the body. The drug has no difficulty in penetrating into the brain and within seconds of inhaling the first puff of marijuana smoke active drug is present on the cannabis receptors in brain. Peak blood levels are reached at about the time that smoking is finished (Fig. 2.5). An experienced marijuana smoker can regulate almost on a puff by puff basis the dose of THC delivered to the brain to achieve the desired psychological effect and to avoid overdose and to minimize the undesired effects. Puff and inhalation volumes tend to be higher at the beginning and lowest at the end of smoking a cigarette (more drug is delivered in the last part of the cigarette because some THC condenses onto this). When experienced smokers were tested with marijuana cigarettes containing different amounts of THC (from 1% to 3.5%), without knowing which was which, they adjusted their smoking behavior to reach about the same level of THC absorption and subjective high. When smoking the less potent cigarettes, puff volumes were larger and puff frequency higher than with the more potent cigarettes, and when smoking the latter, more air was inhaled thereby diluting the marijuana smoke. Many marijuana smokers hold their breath for periods of 10–15

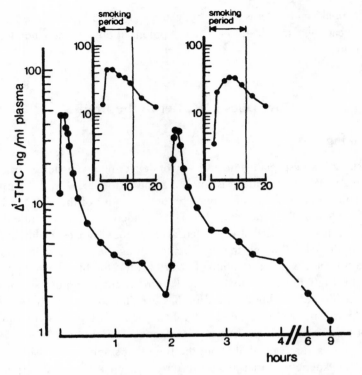

Figure 2.5. Average blood levels of THC in human volunteers who smoked two identical marijuana cigarettes, each containing about 9 mg of THC, 2 hours apart. Insets show the rapid absorption of the drug during the period of smoking. From Agurell et al. (1986), redrawn from L. Hollister et al. (1981) J. Clin. Pharmacol. 21: 1715. Reprinted with permission from Sage Publications Inc.

seconds after inhaling, in the belief that this maximizes the subjective response to the drug. Studies in which subjective responses and THC levels in blood were measured with different breath-hold intervals, however, have failed to show that breath holding makes any real difference to the absorption of the drug—this idea thus seems to fall in the realms of folklore rather than reality.

It is clear why smoking is the preferred route of delivery of cannabis for many people. As with other psychoactive drugs the rapidity by which smoking can deliver active drug to the brain and the accuracy with which the smoker can titrate the dose delivered are powerful pluses. The rapid delivery of the drug to the receptor sites in the brain seems to be an important feature in determining the subjective experience of the high. This is true not only for cannabis but for other psychoactive drugs that are smoked. These include nicotine, crack cocaine, methamphetamine, and increasingly nowadays heroin ("chasing the dragon"). For the narcotic drugs, smoking is the only method that approaches the instant delivery of the drug achieved by intravenous injection — and it does not carry the risks of infection with hepatitis or human immunodeficiency virus (HIV) associated with intravenous use.

Nevertheless, the amount of THC absorbed by smoking varies over quite a large range. Of the total amount of THC in a marijuana cigarette on average about 20% will be absorbed, the rest being lost by combustion, side stream smoke, and incomplete absorption in the lung. But the actual figure ranges from less than 10% to more than 30% even among experienced smokers.

Oral Absorption

Taking THC by mouth is even less reliable as a method of delivering a consistent dose of the drug. THC is absorbed reasonably well from the gut but the process is slow and unpredictable and most of the absorbed drug is rapidly degraded by metabolism in the liver before it reaches the general circulation. The peak blood levels of THC occur anywhere between 1 and 4 hours after ingestion and the overall delivery of active THC to the bloodstream averages less than 10%, with a large range between individuals. The high is correspondingly also delayed by comparison with smoking (Fig. 2.6). Even for the same person the amount of drug absorbed after oral ingestion will vary according to whether they have eaten a meal recently and the amount of fat in their food. A further complication of the oral route is that one of the metabolites formed in the liver is 11-hydroxy-THC (Fig. 2.7). This is a psychoactive metabolite with potency about the same as that of THC. After smoking, the amount of 11-OH-THC formed is relatively small (plasma levels are less than a

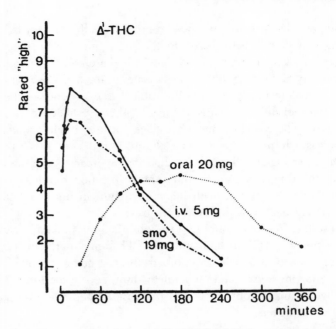

Figure 2.6. Time course of the subjective "high" after administering THC by different routes. Smoking gives as rapid an effect as an intravenous injection, whereas taking the drug by mouth produces a delayed and prolonged "high." The subjective experience somewhat outlasts the presence of THC in blood (c.f. Fig. 2.5) because THC persists longer in the brain. From G. Barnett et al. (1982). J. Pharmacokinetics & Biopharmaceutics 10: 495–506. Reprinted with permission from Plenum Publishing Corp.

third of those for THC), but after the oral route — where all the blood from the intestine must first pass through the liver — the amount of 11-OH-THC in plasma is about equal to that of THC and it probably contributes at least as importantly as THC to the overall effect of the drug.

The only officially approved medicinal formulation of THC (known pharmaceutically as *dronabinol*) is in the form of capsules containing the drug dissolved in sesame oil — a product called Marinol®. It is not surprising that this and other orally administered cannabis products have not proved consistently effective in their medical applications — and both

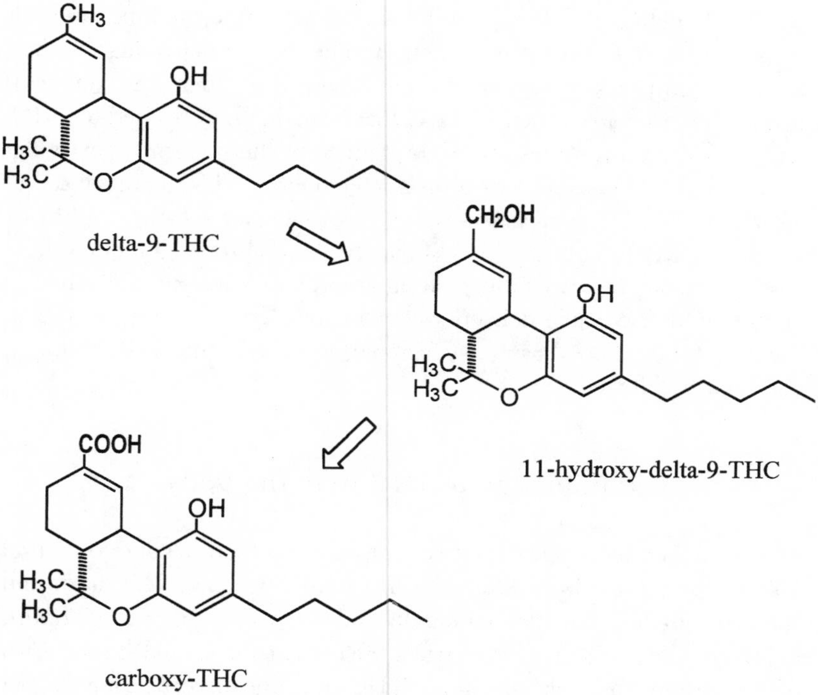

Figure 2.7. Principal route of metabolism of THC.

patients and recreational users generally prefer smoked marijuana. The erratic and unreliable oral absorption of THC poses a serious problem for the effective use of the pure drug as a medicine, as is discussed again in Chapter 4.

Other Routes of Administration

Because THC is so insoluble in water, injection by the intravenous route is very difficult. It can be achieved by slowly adding an alcoholic solution of THC to a rapid intravenous infusion of saline solution, but this is rarely used even in hospital settings. Other alternatives have been little explored so far. By dissolving THC in suitable nontoxic solvents it is possible to deliver the drug as an inhalation aerosol to the lung, and this seems worthy of further examination. Another way of delivering the drug is in the form of a rectal suppository. Research on such a formulation, using a hemisuccinate ester of THC that is gradually converted to THC, has yielded promising results. Absorption from the rectum bypasses the liver and avoids the problem of liver metabolism, which limits the oral availability of THC, and it seems that this route can deliver about twice as much active drug to the bloodstream as the oral route, although there is still considerable variability in drug absorption from one individual to another. Other possible delivery routes include devices designed to heat the drug (or the herbal cannabis material) to vaporize the THC so that it can then be inhaled.

Elimination of THC from the Body

After smoking, blood levels rise very rapidly and then decline to around 10% of the peak values within the first hour (Fig. 2.5). The maximum subjective high is also attained rapidly and persists for about 1 to 2 hours, although some milder psychological effects last for several hours. After oral ingestion the peak for plasma THC and the subjective high is delayed and may occur anywhere from 1 to 4 hours after ingestion, with mild psychological effects persisting for up to 6 hours or more (Fig. 2.6). Although in each case unchanged THC disappears quite rapidly from the circulation, elimination of the drug from the body is in fact quite complex and takes several days. This is largely because the fat-soluble THC and some of its fat-soluble metabolites rapidly leave the blood and enter the fat tissues of the body. As the drug and its metabolites are gradually excreted in the urine (about one-third) and in the feces (about two-thirds) the material in the fat tissues slowly leaks back into the bloodstream and is eventually eliminated. This gives an overall elimination half-time of 3–5 days, and some drug metabolites may persist for several weeks after a single drug exposure (for review see Agurell et al., 1986).

The unusually long persistence of THC in the body has given cause from some concern, but it is not unique to THC — it is seen also with a number of other fat-soluble drugs, including some of the commonly used psychoactive agents, e.g., diazepam (Valium®). The presence of small amounts of THC in fat tissues has no observable effects, as these tissues do not contain any receptors for cannabis. There is no evidence that THC residues persist in the brain, and the slow leakage of THC from fat tissues into blood does not give rise to drug levels that are high enough to cause any psychological effects (Figs. 2.5 and 2.6). Smoking a second marijuana cigarette a couple of hours after the first generates virtually the same plasma levels of THC as previously (Fig. 2.5). Nevertheless, the drug will tend to accumulate in the body if it is used regularly. While this is not likely to be a problem for occasional or light users, there have been few studies of chronic high-dose cannabis users to see whether the increasing amounts of drug accumulating in fat tissues could have harmful consequences. Is it possible, for example, that such residual stores of drug could sometimes give rise to the *flashback* experience that some cannabis users report—the sudden recurrence of a subjective high not associated with drug taking?

The persistence of THC and its metabolites in the body certainly causes confusion in other respects, particularly as drug testing procedures can now detect very small amounts of THC and its metabolites. Urine or blood tests for one of the major metabolites, 11-nor-carboxy-THC (Fig. 2.7), for example, use a very sensitive immunoassay and can give positive results for more than 2 weeks after a single drug exposure. The proportion of the carboxy metabolite relative to unchanged THC increases with time and measurements of this ratio can indicate fairly accurately how long ago cannabis was consumed. The Canadian snowboarder Ross Rebagliati, the first to win an Olympic gold medal for his sport in the 1998 Winter Olympic Games was disqualified the next day because of a positive cannabis test—even though the levels of cannabis and the metabolite measured were barely detectable and it was inconceivable that drug exposure could have affected his performance. Ross adopted the classic defense of "passive smoking" to explain his positive cannabis test (although this would have required Herculean efforts) and the organizers eventually recognized the scientific absurdity of the situation and allowed him to retain his gold medal. For others, being caught with positive cannabis tests, applied randomly in the work place or because they were involved in road traffic accidents or were admitted to hospital emergency rooms the consequences can be more serious.

How Does THC Work?

Pharmacologists used to think that the psychoactive effects of cannabis were somehow related to the ability of the drug to dissolve in the fat-rich membranes of nerve cells and disrupt their function. But the amount of drug that is needed to cause intoxication is exceedingly small. An average marijuana cigarette contains 10–20 milligrams of THC (a milligram is 1/1000 of a gram, or about 1/30,000 of an ounce). Of this, smoking absorbs only 10%–20% — so on average the total body dose is between 1 and 4 mg of THC. The amount of drug ending up in the brain, which accounts for only about 2% of total body weight, can be predicted to be not more than 20–80 micrograms (a microgram is 1/1,000,000 of a gram). Although these are exceedingly small amounts, they are comparable to the naturally occurring amounts of other chemical compounds used in various forms of chemical signaling in the brain. The brain works partly as an electrical machine, transmitting pulses of electrical activity along nerve fibers connecting one nerve cell to another, but the actual transmission of the signal from cell to cell involves the release of pulses of chemical signal molecules known as *neurotransmitters*. These chemicals are specifically recognized by receptors, which are specialized proteins located in the cell membranes of target cells. The neurotransmitter chemicals are released in minute quantities: for example, the total amount of one typical neurotransmitter, noradrenaline, in the human brain is not more than 100–200 micrograms — a quantity comparable to the intoxicating dose of THC. This suggests that THC most likely acts by targeting one or other of the specific chemical signaling systems in brain, rather than by some less specific effect on nerve cell membranes, and indeed this is what the most recent scientific evidence suggests.

An important breakthrough in understanding the target on which THC acts in the brain was the discovery by Allyn Howlett and her col-

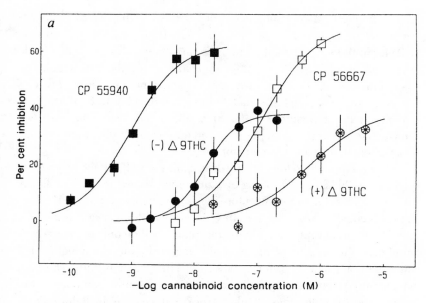

Figure 2.8. Inhibition of cyclic AMP formation in tissue culture cells that possess the CB_1 cannabinoid receptor. The synthetic cannabinoid CP-55940 cs is more potent than (−)delta-9-THC and produces a larger maximum inhibition. The response shows selectivity for the (−) isomers of the compounds versus the (+) isomers (CP-56667 is the (+) isomer of CP-55940). From Matsuda et al. (1990). Reprinted with permission from *Nature*, copyright 1990 Macmillan Magazines Ltd.

leagues at St. Louis University in 1986 of a biochemical model system in which THC and the new synthetic cannabinoid drugs WIN-55,12–2 and CP-55,940 were active (for review see Pertwee, 1995 and Felder and Glass, 1998). The cannabinoids were found to inhibit the activity of an enzyme in the rat brain, adenylate cyclase, which synthesizes a molecule known as cyclic adenosine monophosphate (AMP) (Fig. 2.8). The significance of this finding was that the synthesis of cyclic AMP is known to be controlled by a number of different cell surface receptors that recognize neurotransmitter substances. Some receptors when activated stimulate cyclic

AMP formation, others inhibit it. Cyclic AMP is known as a *second messenger* molecule as it is formed inside cells in response to activation of a receptor at the cell surface by some primary chemical messenger. Cyclic AMP acts as an important control molecule inside the cell, regulating many different aspects of cell metabolism and function. Thus, Howlett's discovery suggested that she had found an indirect way to study drug actions on the cannabis receptor in brain. A few years later in 1988, Howlett's group went one step further and found a more direct way to study drug actions at the cannabis receptor.

A popular method for studying drug actions at cell surface receptors is to measure the specific binding of a substance known to act specifically on such a receptor to the receptor sites in fragments of brain cell membranes incubated in a test tube. In order to be able to measure the very small amounts of drug bound to the receptors — which only occur in small numbers — the drug molecule is usually tagged by incorporating a small amount of radioactivity into the molecule. The radioactive drug can then be measured very sensitively by radioactive detection equipment. Sol Snyder and his colleagues at Johns Hopkins University in Baltimore pioneered the application of this method to the study of drug receptors in brain during the 1970s. In a now famous experiment Snyder and his student Candice Pert used a radioactively labelled derivative of morphine to show that the rat brain possessed specific opiate receptors that selectively bound this and all other pharmacologically active opiate drugs. This experimental approach was subsequently used to devise binding assays for all of the known neurotransmitter receptors in brain and peripheral tissues. These binding assays offer a simple method for determining whether any compound does or does not interact with a given receptor, and provide a precise estimate of its potency by measuring what concentration is needed to displace the radiolabelled tracer.

Snyder's group and several others had tried to see whether a binding assay could be devised for the cannabis receptor by incubating rat brain membranes with radioactively labelled THC in a test tube. This failed, however, because the THC dissolves in the lipid-rich cell membranes very readily and this nonspecific binding to the membranes completely obscured the tiny amount of radiolabelled THC bound specifically to the

Table 2.1. Cannabis Receptor (CB_1) Binding Profiles-[H^3]CP-55,940 Assay Rat Brain Membranes

Drug	Ki—Concentration for Half Occupancy of Receptor Binding Sites—Nanomolar ($10^{-9}M$)
(−)CP-55,940	0.068
(+)CP-55,940	3.4
THC	1.6
11-hydroxy-THC	1.6
Cannabinol	13.0
Cannabidiol	> 500.0

(Devane et al., 1988)

receptors. Howlett collaborated with research scientists at the Pfizer pharmaceutical company to solve this problem. They achieved success by using not THC but the synthetic compound CP-55,940, discovered at the company laboratory as very potent cannabinoid. This had the advantage of being even more potent than THC, and thus binding even more tightly to the cannabis receptor, and as CP-55,940 is more water soluble than THC there was much less nonspecific binding of the radiolabelled drug to the rat brain membrane preparations. The binding assay that resulted seemed faithfully to reflect the known pharmacology of THC and the synthetic cannabinoids (Table 2.1.; Devane et al., 1988). Thus, THC and the psychoactive metabolite 11-hydroxy-THC were able to displace radiolabelled CP-55,940 at very low concentrations — around 1 nanomolar (equivalent to less than 1 microgram in a liter of fluid — and compatible with the amounts of THC thought to be present in the brain after intoxicant doses). Cannabidiol and other inactive cannabinoids were inactive, and the *d*-isomer of CP-55,940 — known to be much less potent in animal behavior models — was some 50 times less potent in displacing the radiolabel than the more active *l*-isomer. The binding assay was quickly adopted and was used, for example, to confirm that the Sterling-Winthrop compound WIN-55,212-2 acted specifically at the cannabis receptor — which most likely explains its analgesic actions. Indeed radioactively labelled WIN-55,212-2 could be used as an alternative label in binding studies to identify the cannabis receptor (Kuster et al., 1993). Another facet of cannabis pharmacology was emphasized by the discovery of these biochemical models — namely that the cannabis receptor seemed to be a wholly novel discovery — not related in any obvious way to any of the previously known receptors for neurotransmitters in the brain. None of the neurotransmitters themselves, or the other chemical modulators in brain, the neuropeptides, interacted to any extent in the cannabis binding assay.

It is notable that the two other naturally occurring cannabinoids cannabidiol and cannabinol interact only weakly with the CB_1 receptor. Nevertheless, these compounds in high doses do possess some pharmacological activities that are probably not related directly to actions on this receptor. Cannabidiol has been reported to be able to protect nerve cells in tissue culture against the toxic effects of L-glutamate (Hampson et al., 1998) and to possess anticonvulsant activity in some animal models of epilepsy (see Chapter 4). In some human psychopharmacology experiments cannabidiol was found to reduce some of the intoxicant effects of THC (Zuardi et al., 1982). The mechanisms involved are unknown, but the effects generally require rather large doses of cannabidiol, suggesting that they may not be of great relevance to understanding how herbal cannabis (that generally contains more THC than cannabidiol) works.

The specific binding of radioactively tagged CP-55,940 could also be used to map the distribution of the receptor sites in the brain. Using thin sections of brain incubated with the labelled compound the location of the binding sites was visualized by overlaying a photographic emulsion sensitive to the radioactive molecule. In this way the distinct pattern of distribution of the cannabis receptor in the brain was described, and found to have the same general pattern in several different mammalian species, including man (Fig. 2.9; Herkenham et al., 1991).

The cannabis receptor in the brain belongs to a family of related receptor proteins, and in 1990 a group working at the United States National Institutes of Health isolated the gene encoding it (Matsuda et al., 1990). This provided independent confirmation of the unique nature of the cannabis receptor. A few years later Sean Munro, in the Medical Research Council Laboratory of Molecular Biology in Cambridge England, discovered a second gene that encoded a similar but distinct subtype of cannabis receptor, now known as the CB-2 receptor, to distinguish it

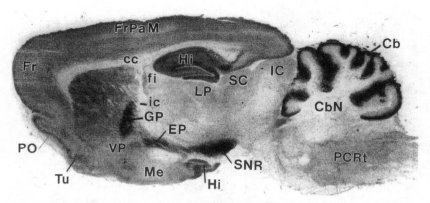

Figure 2.9. Distribution of cannabinoid CB$_1$ receptor in rat brain revealed by an autoradiograph of the binding of radioactively labelled CP-55940 to a brain section. The brain regions labelled are: Cb = cerebellum; CbN = deep cerebellar nucleus; cc = corpus callosum; EP = entopeduncular nucleus; fi = fimbria hippocampus; Fr = frontal cortex; FrPaM = frontoparietal cortex motor area; GP = globus pallidus; Hi = hippocampus; IC = inferior colliculus; LP = lateral posterior thalamus; Me = medial amygdaloid nucleus; PO = primary olfactory cortex; PCRt = parvocellular reticular nucleus; SNR = substantia nigra pars reticulata; Tu = olfactory tubercle; VP = ventroposterior thalamus. Photograph kindly supplied by Dr. Miles Herkenham, National Institute of Mental Health, USA.

from the CB-1 receptor in the brain (for review see Felder and Glass, 1998). CB-2 receptors also bind radioactively tagged CP-55,940 and recognize most of the cannabinoids that act at CB-1 sites. The CB-2 receptor, however, is clearly different and does not occur at all in the brain, being found only in peripheral tissues, particularly on white blood cells — the various components of the immune system of the body. It may be that the actions of THC on such CB-2 sites account for some of the effects of cannabis on the immune system. Research on the CB-2 receptors will be helped by the recent availability of a drug that acts as a selective antagonist at these receptors (SR 144528), and by the development of a genetically modified strain of mice that lack CB-2 receptors.

Cannabinoid Receptor: Agonists, Partial Agonists and Inverse Agonists

Research on the cannabinoid receptor has suggested that THC acts as *partial agonist* at the CB-1 receptor, i.e., it is not able to elicit the full activation of the receptor seen, for example, with the synthetic compounds CP55,940 and WIN55,212-2. This can be shown by the fact that THC does not cause the same maximum inhibition of adenylate cyclase as the synthetic compounds (Fig. 2.8). An alternative functional assay measures the ability of various agonists to stimulate the binding of a metabolically stabilized analogue of GTP (GTP-γ-S) to the activated receptor. In this assay, THC is also only partly effective (25%–30%) by comparison with the synthetic cannabinoids. This assay also reveals some level of constitutive activity in the CB-1 receptor — reflected by binding of GTP-γ-S in the absence of any added cannabinoid. This is inhibited by the antagonist SR141716A, suggesting that in addition to its ability to antagonize the actions of cannabinoid agonists, this compound also acts as an *inverse agonist* at the CB-1 receptor (i.e., it can block the resting level of activity in the receptor that occurs in the absence of cannabinoid agonists). It is not clear whether such constitutive receptor activity is great enough to be of any physiological significance, but if so it might explain some of the pharmacological effects that have been observed in animals when treated with the antagonist alone.

Discovery of Naturally Occurring Cannabinoids

The existence of specific receptors for cannabinoids in brain and in other tissues suggested that they were there for some reason. The receptors have not evolved simply to recognize a psychoactive drug derived from a plant, just as the opiate receptor is not in the brain simply to recognize morphine or heroin. In the 1970s the discovery of the opiate receptor in the brain prompted an intense search for the naturally occurring brain chemicals that might normally activate this receptor — and this revealed the existence of a family of brain peptides known as the *endorphins* (*endo*genous mor-*phines*). Similarly the discovery of the cannabis receptor initiated a search for the naturally occurring cannabinoids (for reviews see Axelrod and Felder, 1998; Piomelli et al., 1998; Felder and Glass, 1998; Di Marzo et al., 1998).

Several laboratories started to work on this problem but the first to come up with an answer was the laboratory of Ralph Mechoulam and his colleagues in Israel, who 30 years earlier had first described THC as the principal active component in cannabis (Devane et al., 1992). The Israelis used pig brain extracts to search for naturally occurring chemicals that could displace the binding of a radioactive cannabinoid in a CB-1 receptor assay in the test-tube, and they focused their attention on chemicals that, like THC, are soluble in fat rather than in water. They succeeded in isolating a tiny amount of a fat derivative, which was active in the test tube receptor assay, and they sent some of this to Roger Pertwee, at the University of Aberdeen in Scotland. He had developed a simple biological assay for THC and related cannabinoids, which involved measuring their ability to inhibit the contraction of a small piece of mouse muscle in an organ bath. The newly isolated chemical was active in these tests — confirming that it had THC-like biological activity. This encouraged the Israeli group to extract a larger amount of material from the pig brain and to determine its chemical structure. It proved to be a derivative of the fatty acid arachidonic acid and they named it *anandamide* after the Sanskrit word *ananda* meaning bliss. Anandamide is a fairly simple chemical, and could readily be synthesized in larger quantities by chemists (Fig. 2.10). There have now been many studies on this cannabinoid, which confirm that it has essentially all of the pharmacological and behavioral actions of THC in various animal models — although when given to animals it is not very potent, because it is rapidly inactivated. Anandamide is present in the brains of all mammals examined so far, including man. A number of synthetic derivatives of anandamide have been prepared and tested, and some have advantages in terms of improved potency and stability, e.g., methanandamide, (Fig. 2.10). Whether this will eventually lead to improved cannabinoids that might be used therapeutically, however, remains to be seen.

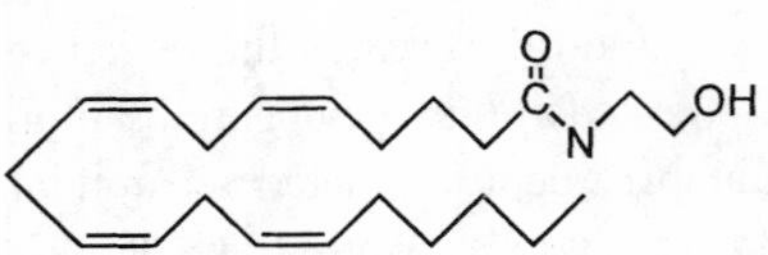

Anandamide 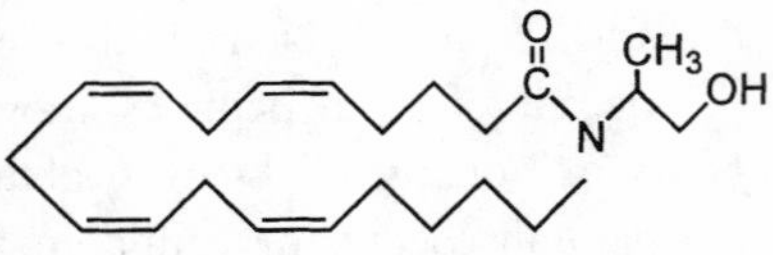   Methanandamide

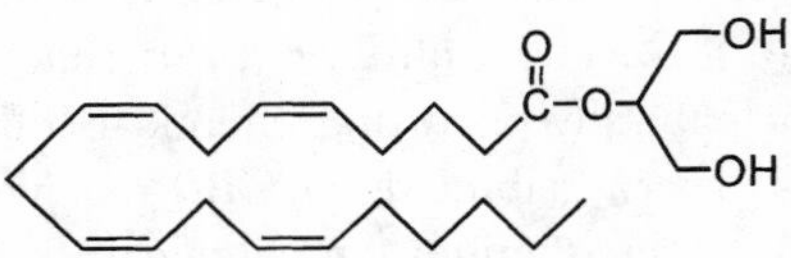

2-Arachidonylglycerol (2-AG)

Figure 2.10. The naturally occurring cannabinoids, anandamide and 2-ararchidonyl glycerol; methanandamide is a synthetic derivative of anandamide that is more resistant to metabolic inactivation.

The discovery of anandamide was not the end of the story. Mechoulam and colleagues (1995) went on to look for anandamide or other cannabinoids in peripheral tissue, using dog intestine as the source. Using a similar approach to that used earlier to identify anandamide they isolated a second naturally occurring cannabinoid, which was also a derivative of arachidonic acid, known as 2-arachidonylglycerol (2-AG) (Fig. 2.10). This too was synthesized and proved to have THC-like actions in various biological tests, including whole animal behavioral models, and potency similar to that of anandamide. It appeared at first that 2-AG might represent the principal natural cannabinoid in peripheral tissues, but Daniele Piomelli and colleagues in California subsequently reported that 2-AG was also present in rat brain where it is far more abundant than anandamide (170 times greater amounts). Since 2-AG is about as potent as anandamide in various biological tests this may suggest that it rather than anandamide is the more important naturally occurring cannabinoid both in brain and periphery. On the other hand, the Piomelli group (Giuffrida et al., 1999) found that they could measure a release of anandamide, but not 2-AG into microprobes inserted into the intact rat brain. It is possible that both anandamide and 2-AG are natural activators of the cannabis receptors in the brain and in the peripheral tissues, with the formation of each compound under separate control. Perhaps one is more important than the other in some parts of the brain or in certain peripheral organs. The details of the natural cannabinoid control system are only just beginning to emerge (Piomelli et al., 1998; Di Marzo et al., 1998).

These discoveries have radically changed the way in which scientists view this field of research. It has changed from a pharmacological study of how the psychoactive drug THC works in the brain to a much broader field of biological research on a unique natural control system, now often referred to as the *cannabinoid system* and the naturally occurring chemicals became known as *endocannnabinoids*. The term *cannabinoid*, originally used to describe the 21-carbon substances found in cannabis plant extracts is now used to define any compound that is specifically recognized by the cannabinoid receptors.

Biosynthesis and Inactivation of Endocannabinoids

Both anandamide and 2-AG are derived from the unsaturated fatty acid arachidonic acid, which is one of the fatty acids found commonly in the phospholipids in cell membranes. It is of interest that arachidonic acid is also a key building block for other groups of lipid chemical messengers, the prostaglandins and leukotrienes. It is not yet clear how anandamide and 2-AG are synthesized in those cells that generate these endocannabinoids. As with the prostaglandins and leukotrienes the endocannabinoids are not stored in cells awaiting release, but rather are synthesized on demand. It is already known that the rate of biosynthesis of anandamideand 2-AG in the brain is increased when nerve cells are activated, for example, by exposure to the excitatory amino acid L-glutamate. Guiffrida et al. (1999) reported the first demonstration of anandamide release from the living brain, using delicate microprobes inserted into a rat brain. They found that anandamide release was stimulated by activation of receptors for the chemical transmitter molecule dopamine, and suggested that this might represent an automatic dampening system, since anandamide seemed to counteract the behavioral stimulant effects of the dopamine-like drug they used. Although 2-AG is present in larger amounts than anandamide, Giuffrida et al. (1999) found no detectable amounts of 2-AG in their release samples.

Although the details of the biosynthetic routes are not yet firmly established it seems likely that 2-AG results from the partial hydrolysis of preexisting arichidonyl membrane lipids, while anandamide may be stored in membrane lipid pools esterified to the third position of phospholipids as N-arachidonylphosphatidylethanolamine (NAPE). An acyl transferase enzymatic step transfers arachidonic acid from the first position of diarachidonyl-glycerophospholipids to the third position of phosphatidylethanolamine to generate NAPE. This is then hydrolyzed by phospholipase D to release anandamide. NAPE can be detected in brain and other tissues in amounts considerably greater than those of anandamide in keeping with its proposed precursor role. Whether specific forms of acyl transferase or phospholipase D are involved in this proposed route of anandamide biosynthesis is not known.

As with other biological messenger molecules, the endocannabinoids are rapidly inactivated after their formation and release. Both anandamide and 2-AG are readily cleaved by hydrolytic enzymes. An enzyme known as fatty acid amide hydrolase (FAAH) seems likely to play a key role. It can hydrolyze both anandamide and 2-AG.

Immunohistochemical staining of rat brain sections using antibodies against purified FAAH showed that the enzyme was most concentrated in regions containing high densities of CB-1 receptors. It was suggested that the degrading enzyme might be located particularly in those neurons that were postsynaptic to axon terminals that bore presynaptic CB-1 receptors The enzyme has been cloned and sequenced; it belongs to the family of serine hydrolytic enzymes. The serine-hydrolase inhibitor phenylmethylsulfonyl fluoride (PSM) inhibits the enzyme and addition of PSM to the incubation fluid increases the affinity of anandamide in binding or functional assays in vitro by a factor of > tenfold, indicating that enzymatic degradation is sufficiently rapid to limit its biological activity. Rapid degradation of anandamide also accounts for its relatively weak and transient actions when administered in vivo. For this reason, a number of metabolically more stable chemical analogues have been synthesized. The simple addition of a methyl group in methanandamide, for example stabilizes the amide linkage and provides a molecule that retains activity at the CB-1 receptor and has greatly enhanced in vivo potency and duration. A number of other stable analogues of both anandamide and 2-AG are now available.

Fatty acid amide hydrolase is an intracellular enzyme and a specific transport protein exists to shuttle anandamide and 2-AG into cells where they can then be metabolized. Similar cellular uptake mechanisms are involved in the inactivation of monoamine and amino acid neurotransmitters and these have become important targets for psychoactive drug development. The antidepressant drug fluoxetine (Prozac®), for example, acts by inhibiting the uptake of serotonin (5-hydroxytryptamine) and making more available to act on serotonin receptors in brain. It is possible that the endocannabinoid transporter and/or the fatty acid amide hydrolase will also become targets for drug discovery. Compounds that selectively interfered with the inactivation of endocannabinoids would represent a novel approach to cannabinoid pharmacology — in contrast to drugs that act directly on cannabinoid receptors such compounds would enhance cannabinoid function only in those tissues where the endogenous system was activated (for review see Piomelli et al., 1998).

Interactions of Cannabinoids with Other Chemical Messenger Systems in the Brain

Although we have only a limited knowledge of how activation of the CB1 receptor in the brain leads to the many actions of THC, some general features of cannabinoid control mechanisms are emerging (for reviews see Pertwee,1995; Piomelli et al., 1998; and Felder and Glass, 1998). Although CB1 receptors are often coupled to inhibition of cyclic AMP formation, this is not always the case. In some nerve cells, activation of CB1 receptors inhibits the function of calcium ion channels, particularly those of the N-subtype. This may help to explain how cannabinoids inhibit the release of neurotransmitters, since these channels are essential for the release of these substances from nerve terminals. CB1 receptors are not usually located on the cell body regions of nerve cells — where they might control the electrical firing of the cells — but are concentrated instead on the terminals of the nerve fibers, at sites where they make contacts (known as synapses) with other nerve cells. Here the CB1 receptors are well placed to modify the amounts of chemical neurotransmitter released from nerve terminals, and thus to modulate the process of synaptic transmission by regulating the amounts of chemical messenger molecules released when the nerve terminal is activated. Experiments with nerve cells in tissue culture, or with thin slices of brain tissue incubated in the test tube have shown that the addition of THC or other cannabinoids can inhibit the stimulation-evoked release of various neurotransmitters, including the inhibitory amino acid γ-aminobutyric acid (GABA), and the amines noradrenaline and acetylcholine. In the peripheral nervous system CB-1 receptors are also found on the terminals of some of the nerves that innervate various smooth muscle tissues. Roger Pertwee and his colleagues in Aberdeen have made use of this in devising a variety of organ bath assays, in which THC and other cannabinoids inhibit the contractions of smooth muscle in the intestine, vas deferens, and urinary bladder, evoked by electrical stimulation. Such bioassays have proved valuable in assessing the agonist/antagonist properties of novel cannabinoid drugs (Pertwee, 1995).

Although the actions of cannabinoids appear generally to be to inhibit neurotransmitter release, this does not mean that their overall effect is always to dampen down activity in neural circuits. For example, reducing the release of the powerful inhibitory chemical GABA might have the opposite effect by reducing the level of inhibition. This may explain two important effects of cannabinoids that have been described in recent years. These are that administration of THC leads to a selective increase in the release of the neurotransmitter dopamine in a region of brain known as the nucleus accumbens and that this is accompanied by an activation and increased release of naturally occurring opioids (endorphins) in the brain (see Chapter 3).

Some Physiological Effects of THC and Anandamide

Heart and Blood Vessels

The cannabinoids exert quite profound effects on the vascular system (Hollister, 1986; Adams and Martin, 1996). In animals, the main effect of THC and anandamide is to cause a lowering of blood pressure; in man the effect in inexperienced users is often an increase in blood pressure but after repeated drug use the predominant effect becomes a lowering in blood pressure. This is due to the action of THC on the smooth muscle in the arteries, causing a relaxation that leads to an increase in their diameter (vasodilatation). This in turn leads to a drop in blood pressure as the resistance to blood flow is decreased, and this automatically triggers an increase in heart rate in an attempt to compensate for the fall in blood pressure. The vasodilatation caused by THC in human subjects is readily seen as a reddening of the eyes caused by the dilated blood vessels in the conjunctiva. The cardiac effects can be quite large — with increases in heart rate in man that can be equivalent to as much as a 60% increase over the resting pulse rate (Fig. 2.11). Although this presents little risk to young healthy people, it could be dangerous for patients who have a history of heart disease, particularly those who have suffered a heart attack or heart failure. Another feature commonly seen after high doses of cannabis is *postural hypotension*, i.e., people are less able to adjust their blood pressure adequately when rising from a seated or lying down position. This leads to a temporary drop in blood pressure, which in turn can cause dizziness or even fainting.

Until recently it was assumed that the effects of the cannabinoids on the heart and blood vessels were mediated indirectly through actions on

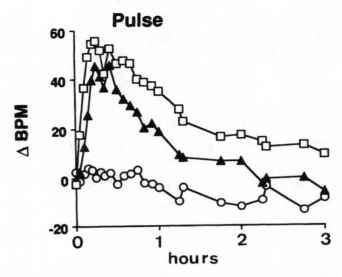

Figure 2.11. Effects on heart rate of smoking a single marijuana cigarette containing 1.75% THC (filled triangles) or 3.55% THC (open squares) versus placebo (open circles). Average results from six volunteer subjects studied on three separate occasions. From Huestis et al. (1992). Reprinted with permission from Masby Inc. St Louis, MO. USA.

receptors in the brain. It is now becoming clear, however, that many and perhaps all of these effects are mediated locally, through CB-1 receptors located in the blood vessels and heart. Isolated blood vessels relax when incubated with anandamide and this effect and the vascular effects in the whole animal can be blocked by the CB-1 antagonist SR141716A. Anandamide is synthesized locally by the endothelial cells lining the blood vessels and it seems to represent an important physiological regulator of the vascular system — along with another potent vasodilator, the gas nitric oxide — also generated locally by endothelial cells.

Other physiological effects of cannabinoids may also be due to direct actions on CB-1 receptors on blood vessels. These include the ability to lower the pressure of fluid in the eyeball (intraocular pressure) — an effect now thought to be due to the presence of CB-1 receptors in the eye; and the ability of cannabinoids to increase blood flow through the kidney—due to a direct action of the drugs on CB-1 receptors on blood vessels in the kidney.

Supression of Immune System Function

Reports during the 1970s seemed at first to provide alarming evidence of a suppression of normal immune system function in chronic marijuana users. Nahas et al. (1974), for example, claimed that white blood cells of the T-cell type isolated from marijuana users and incubated in tissue culture did not show the normal growth and transformation responses when challenged with immune system stimulants. Other reports suggested that T-lymphocytes might be reduced in numbers in marijuana smokers.

Further studies in animals confirmed that treatment with high doses of THC was immunosuppressant. Treated animals were more susceptible to viral or bacterial infections, less able to prevent the growth and spread of small cell cancers, and showed impaired tissue rejection responses, for example, to skin transplants. However, the animal studies required treatment with doses of THC 50–1000 times higher than those taken by human marijuana users, and most would now question the validity of the earlier claims. Several scientists were unable to repeat the early findings, including even Nahas himself. Although there may be some degree of immune system suppression in regular users, particularly in the white blood cells in the lungs of marijuana smokers, there is little evidence that this renders them more susceptible to infection or other disease (Hollister, 1986). Patients suffering from HIV infection might be expected to be at particular risk, since their immune systems are already impaired as a result of the viral infection. However, a longitudinal study involving several thousands of such patients failed to show any effect of marijuana or alcohol use on the progression of the disease to full-blown acquired immunodeficiency syndrome (AIDS) (Kaslow et al., 1989).

Recently there has been renewed interest in the interaction of cannabinoids with the immune system. The second cannabinoid receptor CB-2 is located principally on the various cell types of the immune system, the macrophages, T-cells and B-cells, and mast cells. Although these also express some CB-1 receptors, the CB-2 receptors predominate. It is now also clear that anandamide is synthesized and released by these cells, particularly when challenged with immune system stimulants. Exactly what cellular mechanisms are involved is not known, but exposure of macrophages to anandamide has been shown to stimulate the formation of various cytokines, the chemical mediators of the immune response in such cells. There has been interest by pharmaceutical companies in developing CB-2-selective drugs, which might have utility as immunosuppressants, or in the treatment of such diseases as arthritis or multiple sclerosis, which are thought to be due to inappropriate immune system responses. On the other hand, the recent development of a "CB-2 knock-out mouse," which has been genetically engineered to prevent expression of CB-2 receptors, may dampen this interest. These mice are reported to be remarkably normal, with apparently unimpaired immune system responses and normal numbers of the various white blood cell populations. The precise role of the cannabinoids in modulating immune system function remains to be determined — they clearly represent only one of many complex control mechanisms (for review see Hollister, 1992).

Sex Hormones and Reproduction

This history of research on this topic followed a similar pattern to that described for the immunosuppressant effects of cannabinoids. An initial apparently damning report was followed by a great deal of subsequent research that has largely failed to support the initial claims. A paper published in the prestigious *New England Journal of Medicine* in 1974 sounded the alarm (Kolodny et al., 1974). They reported that blood levels of the male hormone testosterone were severely depressed (average 56% of normal) in 20 young men who were regular marijuana users. In addition, some of the subjects were reported to have reduced sperm counts. These findings, of course, raised immediate concerns about the possibility that marijuana use might impair male sexual function or even lead to impotence. Numerous follow-up studies, however, failed to repeat the original findings with no evidence for altered testosterone levels or anything other than minor reductions in spermatogenesis. Less research has been done in women, although there have been some reports of menstrual cycle abnormalities and transient reductions in prolactin levels. There is no evidence for infertility associated with marijuana use in humans (Hollister, 1986; Zimmer and Morgan, 1997).

Large numbers of animal studies were conducted, and as with the immunosuppressant effects of THC, treatment with high doses was found to lead to consistent effects in suppressing the secretion of both male and female sex hormones, and in female animals this can lead to a temporary suppression of ovulation. Treatment of young animals with THC can also retard adolescent sexual development. Animals, however, develop tolerance to the effects of the cannabinoid and gradually return to normal despite repeated THC dosage. Neither male nor female animals appear to suffer any permanent damage to reproductive function from either acute or chronic THC administration. Nevertheless, cannabinoid CB-1 receptors are present in quite high density in the testes and uterus and anandamide may play some role in sexual function — although this has not yet been clearly defined. One hypothesis is that cannabinoids may be involved in regulating early embryonic development perhaps influencing the window of implantation of the blastocyst in the uterus, but this remains speculative.

Pain Sensitivity

The ability of cannabinoids to reduce pain sensitivity represents an important potential medical application for these substances (see Chapter 4). Animal research on this topic, however, has been hindered by the paucity of animal models that reflect the features of clinical pain and by the lack of selective drugs that act on cannabinoid receptors. The development of such synthetic cannabinoids as CP55940 and WIN55212, which are more water soluble and easier to use than THC, and the availability of selective CB-1 and CB-2 receptor antagonists have strengthened research in this field in recent years (for review see Fields and Meng, 1998). It is clear that cannabinoids are effective in many animal models for both acute pain (mechanical pressure, chemical irritants, noxious heat) and chronic pain (e.g., inflamed joint following injection of inflammatory stimulus or sensitized limb after partial nerve damage). In all these cases the pain-relieving (analgesic) effects of cannabinoids are completely prevented by cotreatment with the CB-1 receptor antagonist SR141716A indicating that the CB-1 receptor plays a key role. In these animal models cannabinoids behave much like morphine, and THC is often found to be approximately equal in potency to morphine. Treatment of animals with low doses of naloxone, a highly selective antagonist of opioid receptor, completely blocks the analgesic effects of morphine but generally has little or no effect in reducing the analgesic actions of THC or other cannabinoids. Conversely, S141716A has little or no effect on morphine analgesia. Thus, the cannabinoids are able to reduce pain sensitivity through a mechanism that is distinct from that used by the opiate analgesics. Nevertheless, there are links between these two systems. In some of the animal models, the analgesic effects of cannabinoids are partially prevented by treatment with naloxone. The cannabinoid antagonist drug SR141716A has also been reported to partially block morphine responses in some studies. In other experiments it has been found that cannabinoids and opiates act synergistically in producing pain relief: i.e., the combination is more effective than either drug alone in a manner that is more than simply additive. For example, in the mouse tail flick response to radiant heat (which measures the time taken to remove the tail from a source of radiant heat) (Smith et al., 1998) and in a rat model of arthritis (inflamed joint) doses of THC that by themselves were ineffective made the animals more sensitive to low doses of morphine. Such synergism could have potentially useful applications in the clinic (Chapter 4).

Another interesting observation is that the CB-1 antagonist SR141716A in addition to blocking the analgesic effects of cannabinoids may sometimes, when given by itself, make animals more sensitive to painful stimuli—i.e., the opposite of analgesia. The simplest interpretation of this finding is that there may be a constant release of endocannabinoids in pain circuits, and that these compounds thus play a physiological role in setting pain thresholds. Alternatively, some of the CB-1 receptors in the body may have some level of activation even when not stimulated by cannabinoids—SR141716A might then act as a so-called inverse agonist to suppress this receptor activity.

It has generally been assumed that the site of action of the cannabinoids in producing pain relief is in the central nervous system (CNS). Indeed CB-1 receptors are known to be present in many of the areas thought to be of key importance in mediating the analgesic effects of opiates. Thus, CB-1 receptors are present in the dorsal horn of spinal cord and in cranial nerve sensory nuclei, areas that receive inputs from peripheral nerves, including those that carry pain information into CNS. Some of the CB-1 receptors may be located on the terminals of primary sensory nerves, where they may control the release of neurotransmitters and neuropeptides involved in the transmission of pain information. In support of the concept of a central site of action, several studies have shown that cannabinoids can produce pain relief in animals when they are injected directly into the spinal cord or brain.

However, some recent research also points to a dual action of the cannabinoids at both CNS and peripheral tissue levels. In a rat inflammatory pain model, in which the irritant substance carrageenan is injected into a paw, two groups have reported that the injection of very small amounts of anandamide into the inflamed paw inhibited the development of increased pain sensitivity normally seen in this model. The effect of anandamide could be prevented by the CB-1 antagonist SR141716A, suggesting an involvement of CB-1 receptors—perhaps located on the peripheral terminals of sensory nerves in the inflamed paw. One group further reported that injection of palmitoylethanolamide, a chemical that is related to anandamide but thought to be selective for CB-2 receptors, also caused analgesia. In this case the CB-2 selective antagonist S144528 blocked the effect but the CB-1 antagonist SR141716A was ineffective. These new findings suggest an important role for peripheral sites in mediating the overall analgesic effects of cannabinoids and point to potential future applications for topically administered cannabinoids in pain control.

Motility and Posture

Cannabinoids cause a complex series of changes in animal motility and posture. At low doses there is a mixture of depressant and stimulatory effects and at higher doses predominantly CNS depression. In small laboratory animals THC and other cannabinoids causes a dose-dependent reduction in their spontaneous running activity. This may be accompanied by sudden bursts of activity in response to sensory stimuli — reflecting a hypersensitivity of reflex activity. Adams and Martin (1996) described the syndrome in mice as follows:

> Δ^9-THC and other psychoactive cannabinoids in mice produce a "popcorn" effect. Groups of mice in an apparently sedate state will jump (hyperreflexia) in response to auditory or tactile stimuli. As animals fall into other animals, they resemble corn popping in a popcorn machine.

At higher doses the animals become immobile, and will remain unmoving for long periods, often in unnatural postures — a phenomenon known as *catalepsy*.

Similar phenomena are observed in large animals. One of the first reports of the pharmacology of cannabis was published in the *British Medical Journal* a hundred years ago (Dixon, 1899). Dixon described the effects of extracts of Indian hemp in cats and dogs as follows:

> Animals after the administration of cannabis by the mouth show symptoms in from three quarters of an hour to an hour and a half. In the preliminary stage cats appear uneasy, they exhibit a liking for the dark, and occasionally utter high pitched cries. Dogs are less easily influenced and the preliminary condition here is one of excitement, the animal rushing wildly about

and barking vigorously. This stage passes insidiously into the second, that of intoxication. . . . In cats the disposition is generally changed showing itself by the animals no longer demonstrating their antipathy to dogs as in the normal condition, but by rubbing up against them whilst constantly purring; similarly a dog which was inclined to be evil-tempered and savage in its normal condition, when under the influence of hemp became docile and affectionate. . . . When standing they hold their legs widely apart and show a peculiar to and fro swaying movement quite characteristic of the condition. The gait is exceedingly awkward, the animal rolling from side to side, lifting its legs unnecessarily high in its attempts to walk, and occasionally falling. A loss of power later becomes apparent especially in the hind limbs, which seem incapable of being extended. Sudden and almost convulsive starts may occur as a result of cutaneous stimulation, or loud noises. The sensory symptoms are not so well defined, but there is a general indifference to position. Dogs placed on their feet will stay thus till forced to move by their ataxia, whilst if placed on their side they continue to lie without attempting a movement. . . . Animals generally become more and more listless and drowsy, losing the peculiar startlings so characteristic in the earlier stage, and eventually sleep three or four hours, after which they may be quite in a normal condition.

Monkeys respond similarly to THC, with an initial period of sluggishness followed by a period of almost complete immobility. The animals typically withdraw into the far corner of the observation cage and adopt a posture that has been called the "thinker position" because the monkeys have a tendency to support their head with one hand and have a typical blank gaze. (Human marijuana users may also sometimes withdraw from contact with other members of the group and remain unmoving for some considerable periods of time).

These effects of cannabinoids most likely reflect their actions on CB-1 receptors in an area of brain known as the basal ganglia, which is importantly involved in the control and initiation of voluntary movements, and a region at the back of the brain known as the cerebellum, which is involved in the fine tuning of voluntary movements and the control of balance and posture. CB-1 receptors are present in some abundance in both of these brain regions (c.f. Fig. 2.9). An intriguing observation is that the CB-1 receptor antagonist SR141716A causes a dose dependent stimulation of running activity in mice. This may suggest that the ongoing release of endocannabinoids in the brain may help to control the level of spontaneous activity in these animals. Alternatively, CB-1 receptors in some regions of their brain could be spontaneously active, and the drug acts as an inverse agonist.

The Billy Martin Tests

As chemical efforts to synthesize novel THC analogues and other synthetic cannabinoids intensified it became increasingly important to have available simple animal tests that might help to predict which compounds retained THC-like CNS pharmacology—in particular which might be psychoactive in man. Although it is never possible to determine whether an animal is experiencing intoxication, certain simple tests do seem to have some predictive value. Professor Billy Martin, who is one of the leading international experts on cannabis pharmacology at the Medical College of Virginia, devised a series of simple behavioral tests that have been used widely (Martin, 1985). He demonstrated that drugs that produced in mice a combination of reduced motility, lowered body temperature, analgesia, and immobility (catalepsy) were very likely to be psychoactive in man. The four symptoms are readily measured experimentally, and exhibit dose-dependent responses to cannabinoids. By testing a large number of compounds Martin and colleagues were able to show that there was a good correlation between the potencies of the various cannabinoids in these tests, and their affinities for the CB-1 receptor, as measured in a radioligand binding assay in the test tube (Fig. 2.12). Furthermore, the CB-1 receptor antagonist SR141716A completely blocks all four responses.

The past 50 years has seen great progress in our understanding of cannabis. This has included the identification of THC as the major psychoactive component in the cannabis plant and the recognition that it acts on specific receptors in the brain and elsewhere in the body. The discovery of the endocannabinoids suggests that THC, like morphine

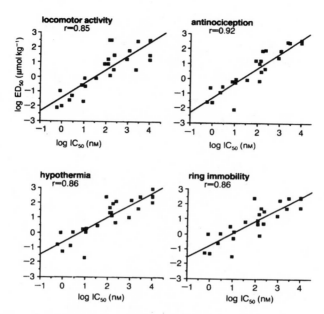

Figure 2.12. The Billy Martin tests. Correlation between *in vivo* and *in vitro* activities of more than 25 cannabinoid analogues to inhibit spontaneous activity ("locomotor activity"), reduce sensitivity to pain (tail-flick test) ("antinociception"), reduce body temperature ("hypothermia"), and cause immobility ("ring immobility") in mice plotted against affinities of the same compounds for CB_1 receptors assessed in an *in vitro* binding assay using radioactively labelled CP-55940. Illustration from Abood and Martin (1992). Reprinted with permission from Elsevier Science.

from the opium poppy, interacts with a naturally occurring regulatory system in the body—the function of which is still largely unknown. The subject of the next chapter, how THC acts on the brain, also remains one of the big questions that has only partly been answered by modern research.

3

The Effects of Cannabis on the
Central Nervous System

A number of approaches can be used to study the effects of drugs on the brain. We can ask people taking the drug to report their own subjective experiences — and there is a large and colorful literature of this type on marijuana. But scientists prefer to use objective methods, and there have been many experiments performed with human volunteers to determine what physiological and psychological alterations in brain function are induced by the drug. The effects that the drug has on animal behavior can also help us to understand how the drug affects the human brain, and understanding how the drug acts in the brain and which brain regions contain the highest densities of drug receptors may also provide useful clues.

Subjective Reports of the Marijuana High

Millions of people take marijuana because of its unique psychotropic effects. It is hard to make a precise scientific description of the state of intoxication caused by marijuana as this is clearly an intensely subjective experience not easily put into words, and the experience varies enormously depending on many variables. Some of these are easily identified:

1. The dose of the drug is clearly important. It will determine whether the user merely becomes *high*, i.e., pleasantly intoxicated, or escalates to the next level of intoxication and becomes *stoned* — a state that may be associated with hallucinations and end with immobility and sleep. High doses of cannabis carry the risk of unpleasant experiences (panic attacks or even psychosis). Experienced users become adept at judging the dose of drug needed to achieve the desired level of intoxication, although this is much more difficult for naive users. The dose is also much easier to control when the drug is smoked, and more difficult when taken by mouth.

2. The subjective experience will depend heavily on the environment in which the drug is taken. The experience of drug taking in the company of friends in pleasant surroundings is likely to be completely different from that elicited by the same dose of the drug administered to volunteer subjects studied under laboratory conditions, or as in some of the

earlier American studies, to convicts in prison who had "volunteered" as experimental subjects.

3. The drug experience will also depend on the mood and personality of the user, their familiarity with cannabis and their expectations of the drug. The same person may experience entirely different responses to the drug depending on whether they are depressed or elated beforehand. Familiarity with the drug means that the user knows what to expect, whereas the inexperienced user may find some of the elements of the drug experience unfamiliar and frightening. The person using the drug for medical reasons has entirely different expectations from those of the recreational user, and commonly finds the intoxicating effects of cannabis disquieting and unpleasant.

There are many detailed descriptions of the marijuana experience in the literature, among the best known are the flowery and often lurid literary accounts of the nineteenth century French authors, Baudelaire, Gautier, and Dumas and those written by the nineteenth century Americans Taylor and Ludlow. Ludlow's book *The Hasheesh Eater* published in 1857 gives one of the best accounts, and will be quoted frequently. Fitz Hugh Ludlow was an intelligent young man who experimented with various mind-altering drugs. He first encountered marijuana at the age of 16 in the local pharmacy, and became fascinated by the drug, and eventually addicted to it. His book vividly describes the cannabis experience, although it is worth bearing in mind that he regularly consumed doses of herbal cannabis extract that would be considered very large by current standards — probably equivalent to several cannabis cigarettes in one session. In modern times there have been several surveys of the experiences of marijuana users. Among these, the books by E. Goode (1970) *The Marijuana Smokers* and by J. Berke and C.H. Hernton (1974) *The Cannabis Experience* which review the 1960s and 1970s experiences of young American and British cannabis users respectively, are particularly useful. Excerpts from Berke and Hernton (1974) are reprinted here with permission from Peter Owen Ltd., London. For "trip reports" from contemporary marijuana users see the web site:
http://www.lycaeum.org/drugs/trip.report

The various stages of the experience can be separated into the *buzz*

leading to the *high* and then the *stoned* states, and finally the *come-down*.
The buzz is a transient stage, which may arrive fairly quickly when smoking.
It is a tingling sensation felt in the body, in the head, and often in the
arms and legs, accompanied by a feeling of dizziness or lightheadedness.

> With hashish a "buzz" is caused, i.e., a tingling sensation forms in the
> head and spreads through the neck and across the shoulders. With a very
> powerful joint this sensation is sometimes "echoed" in the legs.
>
> Usually the first puff doesn't affect me, but the second brings a slight
> feeling of dizziness and I get a real "buzz" on the third. By this I mean a
> sudden wave of something akin to dizziness hits me. It's difficult to de-
> scribe. The best idea I can give is to say that for a moment the whole
> room, people, and sounds around me recede into the distance and I feel as
> if my mind contracted for an instant. When it has passed I feel "normal"
> but a bit "airy-fairy."
>
> (Berke and Hernton, 1974)

During the initial phase of intoxication the user will often experience bodily sensations of warmth (caused by the drug-induced relaxation
of blood vessels and increased blood flow, for example, to the skin). The
increase in heart rate caused by the drug may also be perceived as a
pounding pulse. Marijuana smokers also commonly feel a dryness of the
mouth and throat and may become very thirsty. This may be exacerbated
by the irritant effects of marijuana smoke, but is also experienced when
the drug is taken by mouth.

The influence of the drug on the mind is far-reaching and varied;
the marijuana high is a very complex experience. It is only possible to
highlight some of the common features here. THC has profound effects
on the highest centers in the brain and alters both the manner in which
sensory inputs are normally processed and analyzed and the thinking
process itself. Mental and physical excitement and stimulation usually
accompany the initial stages of the high. The drug is a powerful eupho-
riant, as described so well by Ludlow (1857). Some hours after taking an
extract of cannabis he was

> . . . smitten by the hashish thrill as by a thunderbolt. Though I had felt
> it but once in life before, its sign was as unmistakable as the most famil-
> iar thing of daily life. . . . The nearest resemblance to the feeling is that
> contained in our idea of the instantaneous separation of soul and body.

The hashish high was experienced while Ludlow was walking with a friend, and the effects could be felt during the walk and after they returned home.

> The road along which we walked began slowly to lengthen. The hill over which it disappeared, at the distance of half a mile from me, soon became to be perceived as the boundary of the continent itself. . . . My awakened perceptions drank in this beauty until all sense of fear was banished, and every vein ran flooded with the very wine of delight. Mystery enwrapped me still, but it was the mystery of one who walks in Paradise for the first time. . . . I had no remembrance of having taken hasheesh. The past was the property of another life, and I supposed that all the world was revelling in the same ecstasy as myself. I cast off all restraint; I leaped into the air; I clapped my hands and shouted for joy. . . . I glowed like a new-born soul. The well known landscape lost all of its familiarity, and I was setting out upon a journey of years through heavenly territories, which it had been the longing of my previous lifetime to behold. . . . In my present state of enlarged perception, time had no kaleidoscope for me; nothing grew faint, nothing shifted, nothing changed except my ecstasy, which heightened through interminable degrees to behold the same rose-radiance lighting us up along our immense journey. . . . I went on my way quietly until we again began to be surrounded by the houses of the town. Here the phenomenon of the dual existence once more presented itself. One part of me awoke, while the other continued in perfect hallucination. The awakened portion felt the necessity of keeping in side street on the way home, lest some untimely burst of ecstasy should startle more frequented thoroughfares.

The nineteenth century physician H.C. Wood of Philadelphia described his experimental use of cannabis extract:

> It was not a sensuous feeling, in the ordinary meaning of the term. It did not come from without; it was not connected with any passion or sense. It was simply a feeling of inner joyousness; the heart seemed buoyant beyond all trouble; the whole system felt as though all sense of fatigue were forever banished; the mind gladly ran riot, free constantly to leap from one idea to another, apparently unbound from its ordinary laws. I was disposed to laugh; to make comic gestures.

> (Walton, 1938, p.88)

The initial stages of intoxication are accompanied by a quickening of mental associations and this is reflected typically by a sharpened sense of humor. The most ordinary objects or ideas can become the subjects of fun and amusement, often accompanied by uncontrollable giggling or laughter.

> I often feel very giggly, jokes become even funnier, people's faces become funny and I can laugh with someone else who's stoned just by looking at them.

> I would start telling long involved jokes, but would burst out laughing before completion.

> I nearly always start laughing when in company and have on numerous occasions been helpless with laughter for up to half-an-hour non-stop.
>
> (Berke and Hernton, 1974)

> I get silly. . . . So all kinds of things, like, can crack you up, you know, that aren't really that funny, I guess, in regular life. But they can be really, really funny out of proportion. You can laugh for 20 minutes.
>
> (Goode, 1970)

This effect of the drug is hard to explain, as we know so little about the brain mechanisms involved. Humor and laughter seem to be unique human features. A sharpened sense of humor and increased propensity to laugh are not unique to THC, they are seen with other intoxicants— notably with alcohol. A visit to any lively pub in Britain will confirm this phenomenon. However, THC does seem to be remarkably powerful in inducing a state that has been described as *fatuous euphoria*.

As the level of intoxication progresses from high to stoned (if the dose is sufficiently large) users report feeling relaxed, peaceful, and calm; their senses are heightened and often distorted; they may have apparently profound thoughts and they experience a curious change in their subjective sense of time. As in a dream, the user feels that far more time has passed than in reality it has. As E. Goode puts it:

> Somehow, the drug is attributed with the power to crowd more "seeming" activity into a short period of time. Often nothing will appear to be happening to the outside observer, aside from a few individuals slowly smoking

marijuana, staring into space and, occasionally, giggling at nothing in particular, yet each mind will be crowded with past or imagined events and emotions, and significance of massive proportions will be attributed to the scene, so that activity will be imagined where there is none. Each minute will be imputed with greater significance; a great deal will be thought to have occurred in a short space of time. More time will be conceived of as having taken place. Time, therefore, will be seen as being more drawn out.

(Goode, 1970)

Young British cannabis users report similar experiences:

The strongest feeling I get when I am most stoned is a very confused sense of time. I can start walking across the room and become blank until reaching the other side, and when I think back it seems to have taken hours. Many records seem to last much longer than they should.

Perhaps the "oddest" experience is the confusion of time. One could walk for five minutes and get hung up on something and think that it is an hour later or the other way around, i.e. watch a movie and think it only took five minutes instead of two hours.

(Berke and Hernton, 1974)

Research work at Stanford University in the 1970s by Frederick Melges and colleagues on cannabis users led him to conclude that the disorientation of time sense might represent a key action of the drug, from which many other effects flowed (Melges et al., 1971). His subjects tended to focus on the present to the exclusion of the past or future. Not having a sense of past or future could lead to the sense of depersonalization that many users experience. Focus on the present might also account for a sense of heightened perception, by isolating current experiences from those in the past. This loss of the normal sense of time is probably related to the rush of ideas and sensations experienced during the marijuana high. The user will become unable to maintain a continuous train of thought, and no longer able to hold a conversation.

Sometimes I find it difficult to speak simply because I have so many thoughts on so many different things that I can't get it all out at once.

(Berke and Hernton, 1974)

Perception becomes more sensitive, and the user has a heightened appreciation of everyday experiences. A nurse describes seeing the Chinese-style pagoda in Kew Gardens in London under the influence of marijuana:

> It was like the pagoda had been painted a bright red since I had last seen it—about an hour before. The colour was not just bright, but more than bright, it was a different hue altogether, a deep red, with lots of added pigments, a red that was redder than red. It was a red that leapt out at you, that scintillated and pulsated amid the grey sky of a typical dull English afternoon. Never in a thousand years will I forget that sight. It was like my eyes had opened to colour for the first time. And ever since then, I have been able to appreciate colour more deeply.
>
> (Berke and Hernton, 1974)

New insight and appreciation of works of art have often been reported. Many users report that their appreciation and enjoyment of music is especially enhanced while high; they gain the ability to comprehend the structure of a piece of music, the phrasing, tonalities, and harmonies and the way that they interact. Some musicians believe that their performance is enhanced by marijuana, and this undoubtedly accounted for the popularity of marijuana among jazz players in the United States in the early years of the century. Ludlow described his experience of attending a concert while under the influence of the drug:

> A most singular phenomenon occurred while I was intently listening to the orchestra. Singular, because it seems one of the most striking illustrations I have ever known of the preternatural activity of sense in the hasheesh state, and in an analytic direction. Seated side by side in the middle of the orchestra played two violinists. That they were playing the same part was obvious from their perfect uniformity in bowing; their bows, through the whole piece, rose and fell simultaneously, keeping exactly parallel. A chorus of wind and stringed instruments pealed on both sides of them, and the symphony was as perfect as possible; yet, amid all that harmonious blending, I was able to detect which note came from one violin and which from the other as distinctly as if the violinists had been playing at the distance of a hundred feet apart, and with no other instruments discoursing near them.

While there is no evidence that cannabis is an aphrodisiac it may enhance the pleasure of sex for some people because of their heightened sensitivity and loss of inhibitions. But if the user is not in the mood for sex, getting high by itself will not alter that:

> Hash increases desire when desire is already there, but doesn't create desire out of nothing.

The increased sensitivity to visual inputs tends to make marijuana users favor dimly lit rooms or dark sunshades, as they find bright light unpleasant. The mechanisms in the brain that modulate and filter sensory inputs and set the level of sensitivity clearly become disinhibited. The analysis of sensory inputs by the cerebral cortex also changes in some ways becoming freer ranging — in other ways becoming less efficient. For example, as intoxication becomes more intense, sensory modalities may overlap, so that, for example, sounds are seen as colors, and colors contain music, a phenomenon psychologists refer to as *synesthesia*.

> I have experienced synesthesia — I "saw" the music from an Indian sitar LP. It came in the form of whirling mosaic patterns. I could change the colours at will. At one time a usual facet of a high was that musical sound would take on a transparent crystal, cathedral, spatial quality.
>
> (Berke and Hernton, 1974)

The peak of intoxication may be associated with hallucinations; i.e., seeing and hearing things that are not there. Cannabis does not induce the powerful visual hallucinations that characterize the drug lysergic acid diethylamide (LSD), but fleeting hallucinations can occur, usually in the visual domain.

> . . . occasionally hallucinations. I will see someone who is not there, the much described "insects" which flutter around at the edge of vision, patterns move and swirl.
>
> (Berke and Hernton, 1974)

At the most intense period of the intoxication the user finds difficulty in interacting with others, and tends to withdraw into an introspective state. Thoughts tend to dwell on metaphysical or philosophical topics and the user may experience apparently transcendental insights:

For a single instant, one telling and triumphant moment, I pierced what
Blake might have called "the Mundane Shell." I saw shapes swimming in a
field of neon bands, surging with the colors of Africa. I saw the world
before my eyes through the alchemical crystal revealed, at once, in its
simultaneous complexity and simplicity. My third eye must have blinked.
But only a glimpse — and then, a ripple, a slackening of intensity, and the
moment was lost. . . . This was the most intense visionary experience I
have ever had. And all from a humble green vegetable.

(*http://www.lycaeum.org/drug/trip.report*)
September 7, 1998

The peak period of intoxication is also commonly associated with
daydreams and fantasies.

Fantasies, your thoughts seem to run along on their own to the extent that
you can relax and 'watch' them (rather like an intense day-dream). . . .
Images come to mind that may be funny, curious, interesting in a story-
telling sort of way, or sometimes horrific (according to mood). Also many
other variations.

(Berke and Hernton, 1974)

The nature of the fantasies varies according to personality and
mood. One of the most common fantasies is that of power. The user feels
that he is a god, a superman, that he is indestructible and that all his
desires can be satisfied immediately. Not surprisingly people find such
fantasy states enjoyable and cite them as one of the reasons for their
continued use of the drug. Ludlow described it as follows:

My powers became superhuman; my knowledge covered the universe; my
scope of sight was infinite. . . . All strange things in mind, which had
before been my perplexity, were explained — all vexed questions solved.
The springs of suffering and of joy, the action of the human will, memory,
every complex fact of being, stood forth before me in a clarity of revealing
which would have been the sublimity of happiness.

A curious feature of the cannabis high is that its intensity may vary inter-
mittently during the period of intoxication, with periods of lucidity inter-
vening. There is often the strange feeling of *double consciousness*. Subjects speak of watching themselves undergo the drug-induced delirium,

of being conscious of the condition of their intoxication yet being unable or unwilling to return to a state of normality. Experienced users can train themselves to act normally and may even go to work while intoxicated.

As the effects of the drug gradually wear off there is the "coming down" phase. This may be preceded by a sudden feeling of hunger (*munchies*), often associated with feelings of emptiness in the stomach There is a particular craving for sweet foods and drinks, and an enhanced appreciation and enjoyment of food.

> When I am coming down I generally feel listless and physically weak. . . .
> Often the high ends with a feeling of tiredness, this can be overcome, but
> is usually succumbed to when possible if not by sleep, by a long lay
> down. . . . Conversation initially becomes lively and more intense but as
> the high wears off and everyone becomes sleepy it usually stops. . . .
>
> (Berke and Hernton, 1974)

The cannabis high is often followed by sleep, sometimes with colorful dreams.

However, the cannabis experience is not always pleasant. Inexperienced users in particular may experience unpleasant physical reactions. Nausea is not uncommon, and may be accompanied by vomiting, dizziness, and headache. As users become more experienced they learn to anticipate the wave of lightheadedness and dizziness that are part of the buzz. Even regular users will sometimes have very unpleasant experiences, particularly if they take a larger dose of drug than normal. The reaction is one of intense fear and anxiety, with symptoms resembling those of a panic attack, and sometimes accompanied by physical signs of pallor (the so-called "whitey"), sweating, and shortness of breath. The psychic distress can be intense, as described by young British users:

> I once had what is known as " the horrors" when I had not been smoking
> long. The marijuana was a very strong variety, far stronger than anything I
> had ever smoked before, and I was in an extremely tense and unhappy
> personal situation. I lost all sense of time and place and had slight hallu-
> cinations — the walls came and went, objects and sounds were unreal and
> people looked like monsters. It was hard to breathe and I thought I was
> going to die and that no one would care.

I have felt mentally ill twice when using hashish. On both occasions I felt that I could control no thoughts whatsoever that passed through my mind. It was as though my brain had burst and was distributed around the room. I knew that a short time beforehand I had been quite sane, but that now I was insane and I was desperate because I thought that I would never reach normality again. I saw myself in the mirror, and although I knew that it, the person I saw was me, she appeared to be a complete stranger, and I realized that this was how others must see me. Then the head became estranged from the body — flat piece of cardboard floating a few inches above the shoulders. I was completely horrified, but fascinated, and stood and watched for what must have been some minutes.

(Berke and Hernton, 1974)

As is so often the case, Ludlow's description of a cannabis-induced horror is particularly graphic. After he had taken a much larger dose of cannabis than usual — in the mistaken belief that the preparation was weaker than the one he had used most recently — he went to sleep in a dark room:

. . . I awoke suddenly to find myself in a realm of the most perfect clarity of view, yet terrible with an infinitude of demoniac shadows. Perhaps, I thought, I am still dreaming; but no effort could arouse me from my vision, and I realized that I was wide awake. Yet it was an awaking which, for torture, had no parallel in all the stupendous domain of sleeping incubus. Beside my bed in the centre of the room stood a bier, from whose corners drooped the folds of a heavy pall; outstretched upon it lay in state a most fearful corpse, whose livid face was distorted with the pangs of assassination. The traces of a great agony were frozen into fixedness in the tense position of every muscle, and the nails of the dead man's fingers pierced his palms with the desperate clinch of one who has yielded not without agonizing resistance. . . . I pressed my hands upon my eyeballs till they ached, in intensity of desire to shut out this spectacle; I buried my head in the pillow, that I might not hear that awful laugh of diabolic sarcasm. . . . The stony eyes stared up into my own, and again the maddening peal of fiendish laughter rang close beside my ear. Now I was touched upon all sides by the walls of the terrible press; there came a heavy crush, and I felt all sense blotted out in the darkness.

I awaked at last; the corpse had gone, but I had taken his place upon the bier. In the same attitude which he had kept I lay motionless, conscious, although in darkness, that I wore upon my face the counterpart of his look of agony. The room had grown into a gigantic hall, whose roof was framed of iron arches; the pavement, the walls, the cornice were all of iron. The spiritual essence of the metal seemed to be a combination of cruelty and despair. . . . I suffered from the vision of that iron as from the presence of a giant assassin.

But my senses opened slowly to the perception of still worse presences. By my side there gradually emerged from the sulphurous twilight which bathed the room the most horrible form which the soul could look upon unshattered — a fiend also of iron, white hot and dazzling with the glory of the nether penetralia. A face that was the ferreous incarnation of all imaginations of malice and irony looked on me with a glare, withering from its intense heat, but still more from the unconceived degree of inner wickedness which it symbolized. . . . Beside him another demon, his very twin, was rocking a tremendous cradle framed of bars of iron like all things else, and candescent with as fierce a heat as the fiends.

And now, in a chant of the most terrible blasphemy which it is possible to imagine, or rather of blasphemy so fearful that no human thought has ever conceived it, both the demons broke forth, until I grew intensely wicked merely by hearing it . . . suddenly the nearest fiend, snatching up a pitchfork (also of white hot iron), thrust it into my writhing side, and hurled me shrieking into the fiery cradle. . . .

After more terrible visions Ludlow eventually cried out for help and a friend brought him water and a lamp, upon which his terrors ceased. He was to experience both superhuman joy and superhuman misery from the drug, but became dependent upon it and took it for many years, until after a long struggle he finally gave it up.

Laboratory Studies of Marijuana in Human Volunteers

The sudden popularity of marijuana use among young people in 1960s America prompted an upsurge of scientific research on the drug's effects. A large and often confusing literature emerged, partly because the topic was politically charged from the outset and bias undoubtedly colored some of the investigations. Some researchers seem to have been intent on proving that marijuana was a harmful drug. Others tended to emphasize the benign aspects of the drug.

Studying a psychotropic drug under laboratory conditions is never easy. It is difficult to ensure that subjects receive a standard dose because of the inconsistent absorption of THC — even by regular users. Many of the early studies in the United States used illicit supplies of marijuana of dubious and inconsistent potency. Later standardized marijuana cigarettes became available for academic research studies. They were produced for the National Institute on Drug Abuse, using cannabis plants grown for the government agency by the University of Mississippi. When methods became available for measuring the THC content of the plant material it was possible by judicious blending marijuana of high and low THC content to produce marijuana cigarettes with a consistent THC content. By using plant material with low THC content or marijuana from which THC had been extracted by soaking in alcohol, placebo cigarettes with little or no THC could also be produced.

The question of how to select suitable human subjects for such studies is also difficult. The effects of marijuana in inexperienced or completely naive subjects taking it for the first time are very different from those seen in experienced regular drug users. In one of the very first controlled studies, carried out at Boston University, drug-naive subjects were compared with experienced users. As in many subsequent studies the naive users showed larger drug-induced deficits in the various tasks designed to test cognitive and motor functions than drug-experienced subjects who often show no deficits at all (Weil et al., 1968).

Effects on Psychomotor Function and Driving

Animal experiments have shown that THC has characteristic effects on the ability to maintain normal balance; movements become "clumsy" and at higher doses the animals maintain abnormal postures and may remain immobile for considerable periods. Marijuana similarly affects human subjects, impairing their performance in tests of balance, and reducing their performance in tests that require fine psychomotor control (for example tracking a moving point of light on a screen with a stylus) or manual dexterity. There is a tendency to slower reaction times, although this is a relatively small effect and some studies failed to observe it. In these respects marijuana has similar effects to those observed with intoxicating does of alcohol. An obvious concern is whether these impairments make it unsafe for marijuana users to drive while intoxicated. Driving requires not only a series of motor skills, but also involves a complex series of perceptual and cognitive functions. There have been numerous studies in which the effects of marijuana have been assessed on performance in driving simulators and even a few studies that were conducted in city traffic. Much to everyone's surprise, the results of these studies revealed only relatively small impairments in driving skills, even after quite large doses of the drug. Several of the early studies showed no impairments at all, but as the driving simulators grew more sophisticated and the tasks required more complex and demanding, impairments were observed, for example, in peripheral vision and lane control. Marijuana users, however, seem to be aware that their driving skills may be impaired and they tend to compensate by driving more slowly, keeping some distance away from the vehicle ahead and in general taking less risks. This is in marked contrast to the effects of alcohol, which produces clear impairments in many aspects of driving skill as assessed in driving simulators. Alcohol also tends to encourage people to take greater risks and to drive more aggressively. There is no question that alcohol is a major contributory factor to road traffic accidents; it is implicated in as many as half of all fatal road traffic accidents. Nevertheless, driving while under the influence of marijuana cannot be recommended as safe. Studies in North America and in Europe have found that as many as 10% of the drivers involved in fatal accidents tested positive for THC. However, in a majority of these cases (70%–90%) alcohol was detected as well. It may be that the greatest risk of marijuana in this context is to amplify the impairments caused by alcohol when as often happens both drugs are taken together. Inexperienced drivers and those not accustomed to using marijuana may also be at risk after taking even small doses of the drug.

Flying an aircraft is a more complex task than driving a car, and requires the ability to divide attention between many tasks at once. Flight simulator performance has consistently been found to be impaired after smoking marijuana. In one much cited study of ten experienced licensed private pilots in California, significant deficits in flight simulator performance were observed even 24 hours after a single marijuana cigarette. In one pilot the impairment was sufficiently severe that his aircraft would have landed off the runway! There have been few other reports of such persistent effects and indeed impairments lasting as long as 24 hours are hard to understand since the drug disappears so rapidly from the bloodstream.

Higher Brain Function, Including Learning and Memory

There have been numerous studies of higher brain functions in human subjects given intoxicating doses of marijuana. The results do not always confirm the subjective experiences of the subjects. Thus, while subjectively, users report a heightened sensitivity to auditory and visual stimuli, laboratory tests fail to reveal any changes in their sensory thresholds. If anything they become less sensitive to auditory stimuli. The feeling of heightened sensitivity must, therefore, involve higher perceptual processing centers in the brain, rather than the sensory systems themselves. On the other hand, the perceived changes in the sense of time are readily confirmed by laboratory studies. In one type of test the subject is asked to indicate when a specified interval of time has passed. Intoxicated subjects consistently produce shorter than requested time intervals. In another test the subject is asked to estimate the duration of an interval of time generated by the investigator, in such tests intoxicated subjects overestimate the amount of elapsed time. Thus marijuana makes people experience time as passing more quickly than it really is, or to put it another way marijuana increases the subjective time rate. One minute seems like several.

Many studies have looked for impairments in mental functioning and memory. In simple mental arithmetic tasks, or repetitive visual or auditory tasks that require the subject to remain attentive and vigilant, marijuana seems to have little effect on performance, although if the task requires the subject to maintain concentration over prolonged periods of time (>30 minutes) performance falls off. By far the most consistent and clear-cut acute effect of marijuana is to disrupt short-term memory. Short term memory is usually described as *working memory*. It refers to the system in the brain that is responsible for the short-term maintenance of information needed for the performance of complex tasks the demand planning, comprehension, and reasoning. As described by Baddeley (1996), working memory has three main components: a central executive and two subsidiary short-term memory systems, one concerned with auditory and speech-based information and the other with visuospatial information. These systems hold information and monitor it for possible future use. Working memory can be tested in many ways. In the expanded digit-span test subjects are asked to repeat increasingly longer strings of random numbers both in the order in which they are presented and backwards. In this test marijuana has been reported to produce a dose-dependent impairment in most studies. Other tests involve the presentation of lists of words or other items and subjects are asked to recall the list after a delay of varying interval. Again people intoxicated with marijuana show impairments, and as in the digit-span tests they characteristically exhibit intrusion errors, i.e., they tend to add items to the list that were not there originally. The drug-induced deficits in these tests become even more marked if subjects are exposed to distracting stimuli during the delay interval between presentation and recall. Marijuana makes it difficult for subjects to retain information in working memory in order to process it in any complex manner. The frontal cortex, one of the brain regions that contains a high density of cannabinoid CB-1 receptors, is thought to play a key role in the central executive function, i.e., coordinating information in short-term stores and using it to make decisions or to begin to lay down more stable memories. The hippocampus, another region enriched in cannabinoid receptors, interacts importantly with the cerebral cortex particularly in visuospatial memory and in the processes by which working memory can be converted to longer term storage.

While marijuana has profound effects on working memory, it has little or no effect on the ability to recall accurately previously learned material — it thus seems to have no effect on well-established memories. The relatively severe impairment of working memory may help to explain why during the marijuana high subjects have difficulty in maintaining a coherent train of thought or in maintaining a coherent conversation — they simply cannot remember where the train of thought or the conversation began or the order of the components required to make sense of the information.

The acute effects of marijuana on working memory are relatively short-lived, and disappear after 3–4 hours as the marijuana high wears off. Considerable attention has been paid to the possibility that there might be more persistent effects of marijuana on intellectual function — in particular, whether people who use large doses of marijuana regularly suffer any long-term cognitive impairment. Because of the political implications for marijuana policy, the interpretation of the results of such studies has long been controversial and different studies have sometimes reached apparently divergent conclusions. Fortunately there have been several excellent reviews of this confusing literature, which help to understand it. J.G.C. van Amsterdam et al. (1996), in a report commissioned by the Dutch government point out the many methodological difficulties inherent in studies of the long-term consequences of marijuana use. How does one insure that the results from a group of chronic drug users are compared with a suitable control group of nondrug users, matched for age, educational attainment, and other demographic factors? When should the drug users be tested? Most studies have been done in a period of 12–48 hours after last drug use, but the results may simply reflect a residual effect of the drug, which will continue to leak slowly from fat stores into the blood. Alternatively chronic users may experience withdrawal symptoms when they stop taking marijuana, and this could also impair their cognitive performance during the immediate period after drug cessation. Many of the published studies suffer so severely from such limitations that their conclusions are equivocal at best. Most recent analyses of the literature have concluded that there are indeed significant residual drug effects in the period 12–24 hours after last drug use, and these can be observed in various tests of psychomotor function, attention, and short-term memory. The evidence for any more persistent cognitive deficits is equivocal. Although persistent impairments in various cognitive tests have been reported, these are not consistent from one study to another. The National Institute for Drug Abuse commissioned a series of detailed studies of long-term marijuana users in countries in which heavy use of the drug is common. A series of carefully conducted studies were performed, for example, in Costa Rica, which has a literate westernized culture. Several studies during the 1970s and 1980s compared frequent marijuana users with nonusers using a battery of anthropological and neuropsychological tests, but failed to find any significant differences. It is only recently in a follow-up study reported in 1996 that any significant cognitive differences were found in a cohort of 17 older marijuana users (aged > 45 years). These men had consumed marijuana on average for 34 years, smoking about five joints per day. They were tested after a 72-hour period of abstinence using an impressive array of cognitive tasks designed to investigate various aspects of memory and attention. Statistically significant deficits were observed in only a few of the more complex verbal memory tasks, and these differences were relatively small (less than 10% impairment relative to controls). The same battery of tests applied to a younger group of heavy marijuana users failed to reveal any significant deficits. The authors concluded that:

> . . . the deficiencies observed in this study . . . are subtle. The older long-term users are largely functional and employable, and they do not demonstrate the types of dementia and amnesic syndromes associated with alcohol use of comparable magnitude.
>
> (Fletcher et al., 1996)

Similar studies of long-term heavy users in Jamaica and Greece, countries in which heavy marijuana use is endemic, failed to reveal any notable differences in cognitive function between marijuana users and nonusers. In the 1970s, the United States National Institute of Mental Health commissioned a number of scientific studies to assess the effects of prolonged heavy consumption of cannabis in Jamaica. Comparisons of heavy smokers with nonsmokers revealed surprisingly few adverse effects of smoking on physical health or work performance. In a particularly

famous study Lambros Comitas (1976) reported data that seemed to refute the then popular belief that cannabis consumption led to an
"amotivational syndrome." On the contrary:

> As reflected in their verbal responses, the belief and attitudes of lower class
> users about ganja and work are not at all ambiguous. Ganja is universally
> perceived as an energizer, a motive power — never as an enervator that
> leads to apathy and immobility. In Jamaica, ganja, at least on the ideational
> level, permits its users to face, start and carry out the most difficult and
> distasteful manual labor.
>
> (Comitas, 1976)

Comitas went on to show by objective measurements that the productivity of sugar cane cutters was no different when ganja smokers were
compared with nonsmokers.

Nevertheless, while this may be true for gross deficits in function,
many would now agree that long-term marijuana use can lead to subtle
and selective impairments in cognitive function. This area of research
has been a particular interest of Nadia Solowij, and her recently published monograph *Cannabis and Cognitive Functioning* (Solowij, 1998)
gives an excellent up to date review. Subtle cognitive impairments can be
observed in ex-marijuana smokers in tests that measure the ability to
organize and integrate complex information. The size of the deficit is
related to the frequency of marijuana consumption and the duration. In
addition to deficits in subtle neuropsychological tests, Solowij has described abnormalities in *event related potentials*. These are small electrical discharges that can be recorded from the scalp in response to auditory
stimuli that require the subject to make a decision and take some action.
The electrical potential has a complex waveform, and one component of
this — the so-called P300 was delayed in marijuana users. The results
suggested that subjects were unable to reject complex irrelevant information and hence were less able to focus their attention effectively. In other
words they suffered from a defect in selective attention, a process that is
necessary for the successful completion of most cognitive tasks. Although
these deficits may not have much impact on the ability of ex-marijuana

smokers to function normally they add further weight to the conclusion that marijuana tends to impair executive function in the brain.

Nevertheless, although there have been many rumors that the long-term use of marijuana leads to irreversible damage to higher brain functions the results of numerous scientific studies have failed to confirm this. The report to the Dutch Government prepared by van Amsterdam et al. (1996) sums this up as follows:

> In all studies complete matching of users and non-users was only partly accomplished and the time between cannabis use and testing (duration of abstinence) was too short to ascertain absence of drug residues in the body. Based on the results of the three best studies performed (Schwartz, Pope and Block et al.) residual cognitive effects are seldom observed and if present they are mild in nature.

Comparisons of Marijuana with Alcohol

Alcohol and marijuana are both drugs usually taken in a social context for recreational purposes. Alcohol could be described as the intoxicant for the older generation, marijuana that for the young, although both drugs are quite often consumed together. How do they compare in their effects on the brain? In many ways they are quite similar. A number of studies performed under laboratory conditions have reported that users actually find it difficult to distinguish between the immediate subjective effects of acute intoxication with the two drugs.

Like marijuana, alcohol causes psychomotor impairments, a loss of balance, and a feeling of dizziness or light-headedness. In terms of cognitive performance, both drugs cause impairments in short-term memory while leaving the recall of long-term memories intact. But there are obviously some notable differences. Interestingly the sense of time perception in subjects intoxicated with alcohol is changed in the opposite direction to that observed with marijuana. Tests similar to those described for marijuana above reveal that whereas marijuana speeds up the internal clock alcohol slows it down — 1 minute may seem like several minutes to the marijuana user but feels like only 30 seconds to the alcohol user.

Whereas marijuana tends to make users relaxed and tranquil, alcohol may release aggressive and violent behavior. In terms of the long-term effects of chronic use, alcohol has none of the subtlety of marijuana. Heavy long-term use can lead to organic brain damage and psychosis or dementia (a condition known as Korsakoff's syndrome) while even moderately heavy use can lead to quite severe persistent intellectual impairment.

Where in the Brain Does Marijuana Act?

There have been numerous attempts to identify which brain regions are responsible for mediating the various effects of marijuana. In the 1960s and 1970s many laboratories studied the weak electrical discharges that can be recorded from the scalp as the underlying brain tissue is active — a technique known a electroencephalography (EEG). The results obtained with marijuana, however, were often contradictory — acute reactions to the drug sometimes indicated an activation of the waking type of EEG pattern, but increased slow wave EEG activity characteristic of the resting or sleep state could also sometimes be seen, and there was no obvious localization of the EEG changes to any particular brain region. Measurements of EEG during sleep after marijuana use did show a significant change in sleep EEG patterns. The drug-treated subjects had reduced amounts of the rapid eye movement sleep characteristic of dreaming, and more slow wave sleep, representing the deeper nondreaming state. The EEG is a relatively blunt instrument. It can record only weak electrical signals that arise from the surface of the brain — it cannot give information on changes that may occur in deeper brain structures.

More powerful techniques now exist for monitoring and imaging local changes in brain activity. It is known that changes in the electrical activity of brain cells are closely coupled to changes in regional blood flow in brain. The active nerve cells require more nutrients and oxygen and their activity automatically triggers the dilation of small blood vessels allowing an increase in blood flow. Cerebral blood flow can be monitored in a number of ways. In the most powerful technique, known as

Positron Emission Tomography (PET) a small amount of radioactively labelled water is injected and the radioactive tracer is then imaged by placing the subject in a special camera that detects the low levels of radioactivity in the brain. Alternatively subjects may inhale a small amount of radioactively labelled gas (Xenon). The tracer enters the brain within a few seconds after administration, and the amount present in different brain regions reflects differences in their regional blood flow. The camera images are analyzed by a computerized method known as tomogaphy to yield a three-dimensional image of blood flow throughout the brain. When THC was administered to volunteers with a history of marijuana use there was an increased blood flow in most brain regions, both in the cerebral cortex and in deeper brain structures. These changes reached a peak 30–60 minutes after drug administration, corresponding to the peak period of intoxication. The regions in which the changes in blood flow were greatest and in which there was the best correlation with subjective reports of intoxication were in the frontal cortex (Matthew et al., 1997). This is interesting because it fits well with the concept that frontal cortex is important in the control of "executive" brain functions — which are particularly sensitive to disruption by marijuana (see previous paragraph). Other studies of regional blood flow have reached similar conclusions, with increases in the frontal cortex and temporal cortex reported as the most prominent changes, consistent with the drug-induced impairment of working memory.

Another way of asking which brain regions are involved in the actions of marijuana is to examine the anatomical distribution of the CB-1 receptor in the brain. This has been studied in detail in both animal and human brains. Most of the published reports have used the technique of *autoradiography* to produce images of the CB-1 receptors in thin sections of brain tissue. The tissue sections are incubated with a radioactive cannabinoid (usually CP55,940 containing a radioactive hydrogen atom), as described in Chapter 2, this tracer binds selectively to the CB-1 receptors. Excess radioactive tracer is washed away, and the tissue sections are then covered with a photographic emulsion that is sensitive to radioactivity. After some time the emulsion layer can be processed with a photographic developer and silver grains will become visible in regions that overly areas of the tissue containing CB-1 receptors. In this way a black and white two-dimensional image, known as an autoradiograph, can be created that reflects the distribution of the receptors in that particular section. The differing shades of black and grey will reflect quantitative differences in the densities of receptor sites in different brain areas, and these differences can be measured quantitatively by scanning the auto-radiograph optically with a device that measures differences in light transmission. By examining many different tissue sections, and by cutting these in different planes it is possible to build up a detailed map of the receptor distribution in brain. Examples of such an autoradiograph for the CB-1 receptor in rat brain labelled with ^{3}H-CP55, 940 is illustrated in Figure 2.9. Similar studies have been carried out with sections of human brain obtained postmortem, and the results show that the overall pattern of CB-1 receptor distribution is similar in human brains. In both animals and in man the cerebral cortex, and particularly the frontal regions of cortex, contains high densities of CB-1 sites. These undoubtedly mediate may of the complex effects of marijuana on executive brain functions, e.g., fantasies, depersonalization, and alterations in time sense. There are also very high densities of CB-1 receptor in the basal ganglia, deep brain structures that underlie the cerebral cortex and are involved in the coordination of voluntary movements. The cerebellum at the back of the brain, a region involved in the coordination of balance and fine movements is also rich in CB-1 receptors. The presence of receptors in these two regions probably accounts for the impairments in balance, in walking, and in fine movement control caused by the drug. The pres-ence of CB-1 sites in regions of the "limbic system" known to be impor-tantly involved in emotional behavior may help to explain the eupho-riant effects of the drug, or occasionally its ability to trigger panic/anxiety reactions. Those interested in a more detailed exposition of the receptor topography should read the next paragraph.

Functional Neuroanatomy of Cannabinoid Receptor CB-1 in the Brain

In the cerebral cortex, the density of CB-1 binding sites is almost twice as high in frontal regions as in the posterior occipital cortex (Herkenham et al., 1991). Particularly high receptor densities are found in anterior cingulate cortex, which is one of the regions in which cerebral blood flow studies have shown significant increases correlated to the level of intoxication after marijuana in human subjects. Anterior cingulate is part of the limbic system circuits, which are important in controlling the emotions. In most cortical areas the receptor distribution is laminar, with the highest densities in laminae I and VI. Actions of the drug in the cerebral cortex, particular in the frontal pole, probably account for many aspects of the marijuana high. Cerebral blood flow studies have shown that the strongest correlation of increased cerebral blood flow with intoxication is seen in the right frontal cortex. This is interesting as in most people the right hemisphere is associated with mediation of emotions, while the left hemisphere is more important for analytical thinking and actions. The high densities of CB-1 receptors in basal ganglia are striking. The receptor is especially abundant in the outflow nuclei for striatal efferents, the substantia nigra, and globus pallidus. It seems likely that most of the CB-1 sites in these nuclei are located presynaptically on the terminals of striatal efferents. Excitoxin lesions of the striatum cause a profound loss of CB-1 binding in the globus pallidus and substantia nigra. There is a marked gradient between dorsal regions of the striatum and globus pallidus, which are rich in CB-1 binding sites, and the more ventral regions that have much lower densities of receptors. This is particularly notable in the contrast between dorsal and ventral pallidum. The dorsal regions of striatum are mainly associated with sensory and motor systems and the control of extrapyramidal movements, while the ventral striatum has rich connections with the limbic system and is though to be more involved in the control of motivational and emotional behavior. The distribution of CB-1 sites is, thus, somewhat surprising since although the drug does affect motor function its principal actions are as an intoxicant. The ability of THC to activate dopaminergic activity selectively in regions associated with ventral striatum (for example, the shell region of nucleus accumbens) would also lead one to expect high densities of CB-1 sites in ventral striatum. In the cerebellum, CB-1 receptors are densely present throughout the molecular layer in all parts of the structure, and most are thought to be located presynaptically on the axons and terminals of granule cell fibers. Cerebellar CB-1 receptors probably mediate the ataxia and catalepsy produced by cannabinoids. It is noteworthy that dogs exhibit a prominent ataxia/catalepsy response to cannabinoids, and dog cerebellum is particularly enriched in CB-1 receptors, relative to other mammalian species.

In the hippocampus, CB-1 sites are concentrated on the cell bodies and apical dendrites of pyramidal cells, and these receptors are thought to play an important role in mediating some of the effects of marijuana on short-term memory. There are high densities of CB-1 receptors in the olfactory bulb but these are not clearly associated with any particular aspect of the drug's actions. Receptors are present in some regions of hypothalamus, and these may account for the effects of marijuana on cardiovascular function and the control of body temperature. The relative absence of CB-1 receptors from brainstem structures is undoubtedly related to the low toxicity of the cannabinoids, as they do not directly affect the key brainstem nuclei involved in the control of respiration or blood pressure. Although the density of sites is relatively low, CB-1 receptor binding is present at all levels in the spinal cord, and is enriched in dorsal horn. These sites may be of importance in mediating the analgesic actions of cannabinoids, and in the synergistic interaction of cannabinoids with opioid analgesics.

What Can Animal Behavior Experiments Tell Us?

Studying the actions of psychotropic drugs in animals is inherently difficult — the animals cannot tell us what they are experiencing. The application of ingenious behavioral tests, however, can tell us a great deal about how a drug "feels" to an animal. One technique that is widely used assesses the discriminative stimulus effects of CNS drugs. In this test the animals, usually rats, are trained to press a lever in their cage in order to obtain a food reward, usually a small attractively flavored food pellet, and the reward is given automatically after a certain number of lever presses. The animals are then presented with two alternative levers and must learn to press one (the saline lever) if they had received a saline injection just before the test session, or the other (the drug lever) if they had been injected with the active test drug. Pressing the wrong lever provides no food reward. In other words the animal is being asked, "How do you feel, can you tell that you just received a CNS active drug?" Animals are tested every day for several weeks, receiving drug or saline randomly, and they gradually learn to discriminate the active drug from the placebo (saline). They are judged to have learned the discrimination if they successfully gain a food reward with a minimal number of presses of the wrong lever.

This technique has provided a great deal of valuable information about cannabis and related drugs. Rats and monkeys successfully recognize THC or various synthetic cannabinoids within 2–3 weeks of daily training (Fig. 3.1). The doses of cannabinoids that animals recognize are quite small—less than 1mg/kg orally for THC, and much less for the synthetic cannabinoids WIN 55,212-2 and CP55,940 given subcutaneously (0.032 mg/kg and 0.007 mg/kg respectively; Fig. 3.1; Pério et al., 1996; Torbjörn et al., 1974; Wiley et al., 1995). These doses are in the range known to cause intoxication in human subjects. When animals have been trained to discriminate one of these drugs, the experimenter can substitute a second or third drug and ask the animal another question: "Can you tell the difference between this drug and the one you were previously trained to recognize?" The results of such experiments show that rats and monkeys trained to recognize one of the cannabinoids will generalize (i.e., judge to be the same) to any of the others. They will not generalize, however, to a variety of other CNS-active drugs, including psilocybin, morphine, benzodiazepines, or phencyclidine, suggesting that cannabinoids produce a unique spectrum of CNS effects that the animal can recognize. In all of these studies it was found that the CB-1 receptor-selective antagonist SR141716A completely blocked the effects of the cannabinoids, i.e., when animals are treated with the cannabinoid together with the antagonist they are no longer able to recognize the cannabinoid. These results, thus, provide further strong support for the hypothesis that the CNS effects of THC and other cannabinoids are directly attributable to their actions on the CB-1 receptor in the brain.

Using these techniques one can also ask whether the endogenous cannabinoid anandamide really mimics THC and the other cannabin-

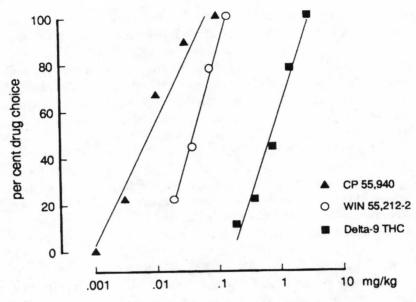

Figure 3.1. Rats trained to discriminate an injection of the synthetic can-nabinoid WIN 55,212-2 (0.3 mg/kg, given subcutaneously) from saline also recognize lower doses of this compound, and the other psychoactive cannabinoids CP-55,940 (given subcutaneously) and THC (given orally). Graph shows percentage of animals selecting the "drug" lever after various doses of the cannabinoids. Results from a group of nine rats. From Pério et al. (1996). Reprinted with permission from Lippincott Williams & Wilkins.

oids. Rats trained to recognize a synthetic cannabinoid do generalize to anandamide, but high doses of anandamide are needed as it is so rapidly inactivated in the body. Monkeys do not generalize to anandamide, prob-ably because it is inactivated too quickly. However, if monkeys are given a synthetic derivative of anandamide that is protected against metabolic inactivation, then they will generalize to this.

In another study rats were trained to recognize THC and were then exposed to cannabis resin smoke. They recognized the cannabis smoke as though it were THC and showed full generalization. In the same study it was found that Δ^9-THC and Δ^8-THC were recognized interchangeably, but there was no generalization between cannabinol or cannabidiol and THC. These results support the hypothesis that THC is the major psychoactive component in cannabis resin and suggest that cannabinol and cannabidiol have little effect.

There is a large literature on the effects of THC and other cannabinoids on various aspects of animal behavior. Unfortunately many studies have used very high doses of THC and the results consequently may have little relevance to how the drug affects the human brain. The human intoxicant dose for THC is less than 0.1 mg/kg, but doses several hundred times higher than this have often been used in animal studies. Such high doses of THC depress most aspects of animal behavior and may cause catalepsy and eventually sleep. Recent work with much smaller doses of cannabinoids has shown the importance of using the appropriate dose. De Fonseca and colleagues (1997) found that low doses of the synthetic cannabinoid HU-210 (0.004 mg/kg) produced behavioral effects in rats suggestive of an antianxiety effect. The test they used was to place the animals in an unfamiliar large open test space containing a dark box to which the animals could retreat. Untreated animals confronted with this novel and unknown environment tend to spend much of their time in the dark box. The animals treated with HU-210, however, appeared to be less fearful and spent more of their time exploring the new environment. If the dose of HU-210 was increased to 0.02 or 0.1 mg/kg a completely different result was obtained, the animals now behaved as thought they were more anxious, and spent most of their time in the dark box. In addition the levels of the stress hormone corticosterone were increased in their blood, suggesting that the high-dose cannabinoid had activated a stress reaction. These findings may have their counterpart in the human experience that low doses of marijuana tend to relieve tension and anxiety, whereas larger doses can sometimes provoke an unpleasant feeling of heightened anxiety or even panic reaction.

In another study, the same group found that administration of low doses of the CB-1 receptor antagonist SR141716A induces anxiety-like effects in the rat, using the same fear-of-novelty type of behavioral tests (Navarro et al., 1997). These findings are very intriguing: they suggest that endogenous cannabinoids in the brain may play a role in fear and anxiety responses and that there is some constant level of activation of CB-1 receptors in the brain by these compounds that can be blocked by SR141716A.

Given the prominent impairment of working memory induced by marijuana in human subjects, it is not surprising that cannabinoids also impair working memory in animals, although there seem to have been rather few such studies. In animals there are a number of ways of assessing working memory. One model much used in rodents is the radial maze. In this a rat or mouse is placed at the center of a maze with eight arms projecting away from the central area. At the start of each experiment all eight arms contain a food reward. The animal is placed at the center of the maze and enters one arm to retrieve a food reward. The animal is then returned to the central area and all eight arms are temporarily blocked by sliding doors. After a delay, usually of only a few seconds, the doors are opened again and the animal is free to retrieve more food rewards. Success depends on being able to remember which arms had already been visited, to avoid fruitless quests. After daily training for 2–3 weeks the animals become quite expert at the task and retrieve all eight food rewards while making few errors. THC and other cannabinoids will disrupt the behavior of such trained animals in a dose-dependent manner. Furthermore, this effect of the cannabinoids can be prevented by SR141716A showing that it is due to an action of THC on CB-1 receptors. The synthetic cannabinoids CP55,940 and WIN 55,212-02 are also effective in this model and they are considerably more potent than THC. CP55,940 will also disrupt this behavior when injected in minute amounts directly into the rat hippocampus, a structure known to be particularly important for spatial memory.

The cannabinoids have also been shown to disrupt the phenomenon of long-term potentiation in the hippocampus. In this model, slices of rat hippocampus are incubated in saline and electrical activity recorded from nerve cells by miniature electrodes. A burst of electrical stimulation of input nerve pathways to the hippocampus leads to a long lasting potentiation, so that further periods of less intense stimulation lead to greater responses than previously. This form of plasticity in neural circuits is thought to be critical in the laying down of memory circuits in the brain. When cannabinoids are added to the incubating solution they disrupt this potentiation.

Another behavioral test that can be employed both in rodents or in monkeys is the *delayed matching to sample* task. When using this test in monkeys an animal is confronted with a number of alternative panels on a touch screen. At the start of the experiment one of these panels is illuminated and the screen then goes dead, preventing the animal from making any immediate response. After a delay, usually of 30–90 seconds all the panels on the screen are illuminated and the animal has to remember which one was illuminated earlier and press it to obtain a food reward. After daily training sessions animals become proficient at such tasks and make few errors. THC and other cannabinoids again disrupt behavior in these tests of working memory. Similar results have been observed in rats using a variant of this task.

The results of a recent study suggest the possibility that the ongoing release of endogenous cannabinoids in the brain may play a role in modulating working memory; the study employed an unusual memory task involving a social recognition. When adult rats or mice are exposed for the first time to a juvenile animal they spend some time contacting and investigating it. If the adult is exposed to the same juvenile within 1 hour of the first encounter it appears to recognize that it has already encountered this juvenile and will spend less time investigating it. If the delay between trials is increased to 2 hours, however, the adult seems to have largely forgotten the original encounter and investigates the juvenile animal thoroughly once more. This short-term memory appears to rely mainly on olfactory cues. Researchers at the Sanofi company in France found that animals treated with low doses to the cannabinoid antagonist SR141716A showed improved memory function in this test, and were able to retain the social recognition cues for 2 hours or more. They also showed that the performance of aged rats, who had difficulty in remembering for even as long as 45 minutes, could be significantly improved by treatment with the antagonist drug (Terranova al., 1996). This raises the intriguing possibility that cannabinoid receptor antagonists could possibly have beneficial effects in elderly patients who suffer from memory loss.

The powerful effects of the cannabinoid drugs in the test may be related to the fact that social recognition in rodents importantly involves olfactory cues, and the CB1 receptor is present in especially high densities in the olfactory regions of the brain.

Does Repeated Use of Marijuana Lead to Tolerance and Dependence?

Many drugs when given repeatedly tend to become less and less effective so that larger and larger doses have to be given to achieve the same effect, i.e., *tolerance* develops. There are many examples of tolerance to THC and other cannabinoids in animals treated repeatedly with these drugs. This literature was well reviewed by Pertwee (1991). Tolerance can be seen even after treatment with quite modest doses of THC, but is most profound when large doses (>5 mg/kg) are employed. With very high doses (as much as 20 mg/kg per day) animals may become almost completely insensitive to further treatment with THC. When animals become tolerant to THC they also demonstrate cross-tolerance to any of the other cannabinoids, including the synthetic compounds WIN 55,212-2 and CP 55,940. This suggests that the mechanism underlying the development of tolerance has something to do with the sensitivity of the cannabinoid receptors or some mechanism downstream of these receptors, rather than simply to a more rapid metabolism or elimination of the THC. Repeated treatment with THC in both animals and people does tend to lead to an increased rate of metabolism of the drug — probably because drug-metabolizing enzymes in the liver are induced by repeated exposure to the drug. But these changes are not big enough to explain the much larger changes in sensitivity seen in responses to the drug — these include effects on cardiovascular system, body temperature, and behavioral responses. A more likely explanation is that repeated exposure to high doses of THC leads to a compensatory decrease in the sensitivity or number of cannabinoid receptors in brain. Decreases in the density of CB-1 receptor binding sites have been demonstrated experimentally in the brains of rats treated for 2 weeks with high doses of THC or CP 55,940.

In human volunteers exposed repeatedly to large doses of THC under laboratory conditions, tolerance to the cardiovascular and psychic effects can be produced as in the animal studies. However, it is not clear that tolerance occurs to any significant extent in people who use modest amounts of marijuana. The casual user, taking the drug infrequently or those using small amounts for medical purposes seem to develop little if any tolerance. Tolerance seems only likely to become important for heavy users who are taking large amounts on a daily basis. Even for such people firm evidence that tolerance becomes an important factor is lacking. Indeed there have been reports that heavy users can become sensitized, so that even rather small doses of the drug can send them into a high. This was described by Ludlow (1875):

> Unlike all other stimuli with which I am acquainted, hasheesh, instead of requiring to be increased in quantity as existence in its use proceeds, demands rather a diminution, seeming to leave, at the return of the natural state . . . an unconsumed capital of exaltation for the next indulgence to set up business upon.

This might have been literally true, as Ludlow consumed such large doses of cannabis extract that there could have been a significant accumulation of the drug in his body, so that subsequent doses acted upon a preexisting baseline. Those like Ludlow who consume the drug by mouth are also more likely to induce the synthesis of additional amounts of the enzymes in the liver, which metabolizes the drug. Although most of this metabolism leads to the formation of inactive byproducts, one metabolite formed in the liver, 11-hydroxy-THC, is even more psychoactive than THC itself. This may be formed in unusually large amounts in regular heavy users of orally administered cannabis.

Whether tolerance or sensitization develops to the repeated use of cannabis probably depends both on the route by which the drug is taken and the quantity. Surveys of recreational cannabis users in Britain, described in the House of Lords Report (1998), show that the quantities of drug consumed vary widely. *Casual* users take the drug irregularly, in amounts of up to 1 g of resin at a time, with an annual total of not more than 28 grams (g) (1 ounce). *Regular* users typically consume 0.5 g of resin a day (equivalent to 3 or 4 smokes of a joint or pipe), while *heavy* users often consume more than 3.5 g of resin a day, more than 28 g a week. Such heavy users will be more or less permanently stoned. Many heavy users report tolerance to the drug, and they may require 5–10 times higher doses than casual users in order to become high. In some countries where the use of marijuana is endemic, very large amounts of the drug are regularly consumed.

The question of whether regular users become dependent on the drug has proved to be one of the most contentious in the whole field of cannabis research. Those opposed to the use of marijuana believe that it is a dangerous drug of addiction, by which young people can easily become hooked. On the other hand, proponents of cannabis claim that it does not cause addiction and dependence at all, and users can stop at any time of the own free will. To understand these opposing views it is important to be clear what we mean when we use the terms tolerance, addiction, and dependence. As the House of Lords Report (1998) puts it:

The consumption of any psychotropic drug, legal or illegal, can be thought of as comprising three stages: use, abuse and addiction. Each stage is marked by higher levels of drug use and increasingly serious consequences.

Abuse and addiction have been defined and redefined by various organisations over the years. The most influential current system of diagnosis is that published by the American Psychiatric Association (DSM-IV, 1994). This uses the term "substance dependence" instead of addiction, and defines this as a cluster of symptoms indicating that the individual continues to use the substance despite significant substance-related problems. The symptoms may include "tolerance" (the need to take larger and larger doses of the substance to achieve the desired effect), and "physical dependence" (an altered physical state induced by the substance which produces physical "withdrawal symptoms," such as nausea, vomiting, seizures and headache, when substance use is terminated); but neither of these is necessary or sufficient for the diagnosis of substance dependence. Using DSM-IV, dependence can be defined in some instances entirely in terms of "psychological dependence"; this differs from earlier thinking on these concepts, which tended to equate addiction with physical dependence."

For details of the DSM-IV criteria see the box below.

DSM-IV Criteria for Substance Dependence
(American Psychiatric Association, 1994)

A maladaptive pattern of substance abuse, leading to clinically significant impairment or distress, as manifested by three (or more) of the following, occurring at any time in the same 12-month period:

1. Tolerance, as defined by either of the following:
   (a) A need for markedly increased amount of the substance to achieve intoxication or desired effect.
   (b) Markedly diminished effect with continued use of the same amount of the substance.
2. Withdrawal, as defined by either of the following:
   (a) The characteristic withdrawal syndrome for the substance
   (b) The same (or a closely related) substance is taken to relieve or avoid withdrawal symptoms.
3. The substance is often taken in larger amounts or over a longer period than was intended.
4. There is a persistent desire or unsuccessful efforts to cut down or control substance use.
5. A great deal of time is spent in activities necessary to obtain the substance (e.g., visiting multiple doctors or driving long distances), use the substance (e.g., chain-smoking), or recover from its effects.
6. Important social, occupational, or recreational activities are given up or reduced because of substance use.
7. The substance use is continued despite knowledge of having a persistent or recurrent physical or psychological problem that is likely to have been caused or exacerbated by the substance (e.g., current cocaine use despite recognition of cocaine-induced depression or continued drinking

despite recognition that an ulcer was made worse by alcohol consumption).

Substance abuse with physiological dependence is diagnosed if there is evidence of tolerance or withdrawal.
Substance abuse without physiological dependence is diagnosed if there is no evidence of tolerance or withdrawal.

(American Psychiatric Association, 1994)

This new way of thinking about drug dependence is significantly different from much of the earlier work in this field. It means that neither tolerance nor physical dependence need necessarily be present to make the diagnosis of substance dependence. This has particularly changed the way in which cannabis is currently viewed. It has often been argued that since neither tolerance nor physical dependence are prominent features of regular marijuana users that therefore the drug cannot be addictive. The DSM-IV definition of *substance dependence* is made as the result of a carefully structured interview, and the diagnosis rests on the presence or absence of various items from a checklist of symptoms. When such assessments are made on groups of regular marijuana users a surprisingly high proportion are diagnosed as dependent. Dr. Wayne Hall and his colleagues in Australia, for example, recently reported the results of studies of this type in people who had been regular heavy users of marijuana for several years — as many as 50% of them were diagnosed as dependent. Although a recent WHO report (1997) on cannabis predicted that as many as half of all those who use marijuana daily will become dependent it is likely that this is an overestimate. The groups studied in Australia were also particularly heavy users. It is hard to estimate just what proportion of regular cannabis users will become dependent. There will be varying levels of dependence, and this will undoubtedly be influenced by the amount of drug consumed and for how long. In some people the drug will come to dominate their lives. They will feel a psychological need and craving for the drug, and will become preoccupied with locating continuing supplies of the drug. Consumption of marijuana may become so frequent that the user may remain almost permanently stoned. They may prepare a joint before going to sleep at night in order to ensure that it is available for the morning. The severely dependent user is permanently cognitively impaired, lacks motivation, tends to suffer from lowered self-esteem and may be depressed, and is unlikely to be able to function at all in work or education. Although most regular cannabis users suffer merely mild discomfort when they stop taking the drug, the severely dependent user will suffer a definite syndrome of unpleasant withdrawal symptoms — including, anxiety, depression, sleep disturbance, nausea, and loose stools or diarrhea. The phenomenon was well recognized by Ludlow more than a hundred years ago, since as a result of regular ingestion of large doses of herbal cannabis extract he himself became dependent on the drug. His book describes graphically his struggle to end the habit — which he eventually succeeded in doing. He initially regarded his experiences with cannabis in the nature of a scientific experiment, but he describes his first recognition of the problem:

> At what precise time in my experience I began to doubt the drug being, with me, so much a mere experiment as a fascinating indulgence, I do not now recollect. It may be that the fact of its ascendancy gradually dawned upon me; but at any rate, whenever the suspicion became definite, I dismissed it by so varying the manner of the enjoyment as to persuade myself that it was experimental still.

Ludlow continued to take the drug, but he experienced increasingly terrifying fantasies and hallucinations. He determined to reduce the dose gradually, but was not very successful:

> The utmost that could be done was to keep the bolus from exceeding fifteen grains. From ten and five, which at times I tried, there was an insensible sliding back to the larger allowance, and even there my mind rebelled at the restriction.

Eventually by force of will he stopped taking cannabis altogether, and experienced an unpleasant period of withdrawal, withdrawing into himself and avoiding social contact, suffering from depression, and experiencing hashish-like dreams:

My troubles were not merely negative, simply regrets for something which
was lost, but a loathing, a fear, a hate of something which was. The very
essence of the outer world seemed a base mockery, a cruel sham.

It is becoming increasingly clear that cannabis is a drug to which
regular users can become dependent, and that this adversely affects large
numbers of people. Cannabis dependence is still largely unrecognized,
because it is still widely believed that it is not an addictive drug. Less
than half of the Australian cannabis users diagnosed as dependent by
DSM-IV criteria in Dr. Hall's studies were willing to admit that they
were dependent on the drug. There is a real need to educate cannabis
users, to convey the message that they do run a risk of allowing the drug
to dominate their lives.

On the other hand, if one attempts to assess this risk by comparison
to other addictive drugs cannabis does not score top of the list in terms of
either the severity of the addiction or the likelihood of becoming hooked.
Cocaine and heroin are far more damaging, both in terms of the severity
of the physical withdrawal syndrome that users will experience if they
stop taking the drug and in the probability of becoming hooked on the
drug. Nicotine is notorious in the sense that very high proportions of
cigarette smokers tend to become permanent smokers after consuming
only a few packets of cigarettes. Unlike the casual user of marijuana the
cigarette smoker typically smokes 15–20 cigarettes a day every day of the
year. Unlike cigarette smokers, most marijuana smokers also seem to be
able to give up the habit relatively easily. As they reach their 30s and
become responsible for a family and a job they are no longer willing to
take the risk of being punished for illegal drug use. Most marijuana users
in Europe and the United States are people in their late teens and twenties, with relatively few over the age of 30.

There have also been developments in basic research that point to
similarities between cannabis and other drugs of addiction. The availability of the CB-1 receptor antagonist SR141716A, for example, has shown
that physical dependence accompanied by a withdrawal syndrome can
be seen in animals that have been treated for some time repeatedly with
THC or other cannabinoid when they are challenged with the antagonist

drug. The withdrawal signs in rats, for example, included *wet-dog shakes* (a characteristic convulsive shaking of the body as though the animal's fur were wet—a behavior also seen typically during opiate withdrawal), scratching and rubbing of the face, compulsive grooming, arched back, head shakes, spasms, and backwards walking. In dogs, the withdrawal signs included withdrawal from human contact, restlessness, shaking and trembling, vomiting, diarrhea, and excess salivation. The reason such withdrawal signs are not normally seen in animals or in people when cannabinoid administration is suddenly stopped is probably related to the long half-life of THC and some of its active metabolites in the body. This means that the CB-1 receptor is still exposed to low levels of cannabinoid for some time after the drug is stopped. With the antagonist drug, however, the CB-1 receptor is suddenly blocked. These findings have an interesting parallel with research on the benzodiazepine tranquilizers, of which Valium® (diazepam) is the best known example. These too were thought not be addictive, since there was little evidence for any withdrawal syndrome on terminating drug treatment. When the first benzodiazepine receptor antagonist drug flumazenil became available though, it soon became clear that withdrawal signs could be precipitated in drug-treated animals when challenged with this antagonist. As with THC, the benzodiazepines persist for long periods in the body so drug withdrawal can never be abrupt. It is now generally recognized that benzodiazepine tranquilizers and sleeping pills do carry a significant risk of dependence on repeated use.

One way in which scientists can assess the addictive potential of psychoactive drugs is to see whether animals can be trained to self-administer them. Self-administration of heroin or cocaine is easily learned by rats, mice, or monkeys. Indeed rats will self-administer cocaine to the exclusion of all other behavior, including feeding and sex. They have to be given restricted access to the drug to avoid damaging their health. It has proved very difficult or impossible to train animals to self-administer THC, however, and this has often been used to argue that THC has no addictive liability. But THC is very difficult to administer to animals because of its extreme insolubility, which precludes intravenous injection, the preferred route for giving addictive drugs. Recently though it has been shown that mice can learn to self-administer the more water-soluble cannabinoid WIN 55, 2212–2 (Ledent et al., 1999).

Another series of experiments in animals has revealed that in common with all other drugs of addiction, THC is able to selectively activate nerve cells in the brain that contain the chemical transmitter dopamine. French et al. (1997) in Arizona first reported that small doses of THC activated the electrical discharge of dopamine-containing nerve cells in the ventral tegmentum region of rat brain—which they recorded electrically with microelectrodes. Tanda et al. (1997), working in Sardinia, subsequently confirmed this by direct measurements of dopamine release from the nucleus accumbens region of the rat brain, which contains the terminals of the nerves originating from the ventral tegmentum (Fig. 3.2). They perfected a delicate technique that involves the insertion of minute probes into this region of rat brain, through which chemicals released in the brain can be monitored continuously in conscious freely moving animals (a method known as microdialysis). Earlier work from this group and a number of other laboratories had shown that a number of drugs of addiction selectively activate dopamine release in this region of the brain, the drugs included heroin, cocaine, d-amphetamine, and nicotine. To this list they now added THC, adding to speculation about its status as a drug of addiction. Furthermore, the Italian group reported

---

Figure 3.2. Release of dopamine from intact rat brain measured using microdialysis probes. *A:* Dopamine release is stimulated by the administration of THC (0.15 mg/kg, i.v.) or heroin (0.03 mg/kg, i.v) (circles). Filled circles indicate data points that were significantly different from baseline control values. In animals treated with the opiate μ receptor antagonist naloxonazine, neither THC nor heroin caused dopamine release any longer (squares). *B:* Sections of rat brain drawn to indicate the positions of the microdialysis probes in the individual animals used; Core = core of nucleus accumbens; Shell = shell of nucleus accumbens; CPu = caudate putamen; SN = substantia nigra; VTA = ventral tegmentum. On each section **A** indicates the anterior coordinate, measured in millimetres from bregma. From Tanda et al. (1997). Reprinted with permission from *Science*. Copyright 1997 American Association for the Advancement of Science.

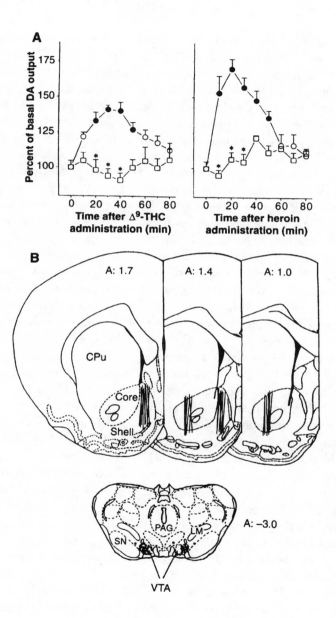

that the THC-induced release of dopamine seemed to involve an opioid mechanism — since the effect of THC could be prevented by treatment of the animals with naloxonazine, a drug which potently and selectively blocks opioid receptor sites in the brain. These results thus suggested that THC acts in part by promoting the release of opioid peptides in certain regions of the brain, and that one of the consequences of this is to cause an increase in dopamine release in the nucleus accumbens. The precise biological meaning of this remains unclear. Most scientists do not believe that dopamine release per se explains the pleasurable effects of drugs of addiction — but it does seem to have some relation to whether the animal or person will seek to obtain further doses of that drug. Dopamine release in the nucleus accumbens is triggered by a variety of stimuli that are of significance to the animal — including food and sex. The ability of THC to activate opioid mechanisms also does not mean that THC is equivalent to heroin. Clearly animals and humans can readily distinguish the distinct subjective experiences elicited by the two drugs, and THC or other cannabinoids do not mimic the severe physical dependence and withdrawal signs associated with chronic heroin use. Nevertheless, there is growing evidence that the naturally occurring opioid and cannabinoid systems represent parallel and sometimes overlapping mechanisms. Rats made dependent on heroin and then challenged with the opiate antagonist naloxone exhibit a strong withdrawal syndrome, with various characteristic behavioral features — for example, wet-dog shakes, teeth chattering, writhing, jumping, and diarrhea. Interestingly some of these features are seen in a milder form if heroin-dependent animals are challenged with the cannabinoid antagonist SR141716A. Conversely rats treated repeatedly with high doses of cannabinoids will exhibit mild signs of withdrawal when challenged with the opiate antagonist naloxone. More support for the concept of a link between the cannabinoid and opioid systems in brain has come from a recent report on a new strain of genetically engineered mice that lack the cannabinoid CB1 receptor (Ledent et al., 1999). These animals survive quite normally without the CB1 receptor, but as expected they are unable to show any of the normal CNS responses to THC (analgesia, sedation, and hypothermia). Interestingly, the mice were also less responsive to morphine. Although morphine was still analgesic, it was less likely to be self-administered, and the mice displayed a milder opiate withdrawal syndrome.

Further support for the existence of a genuine cannabis withdrawal syndrome in animals came from De Fonseca et al. (1997) who reported that there were elevated levels of the stress-related chemical corticotropin releasing factor (CRF) in rat brain when rats were withdrawn from treatment for 2 weeks with the potent cannabinoid HU-210. Elevated levels of brain CRF were also seen in animals during withdrawal from alcohol, cocaine, and heroin. The association of withdrawal with unpleasant anxiety and stress reactions is perhaps one reason people continue to use drugs of dependence.

The new evidence shows more clearly than before that the repeated dosing with cannabis can lead to dependence and withdrawal in animals, and that these phenomena resemble those seen after treatment with other drugs that possess addictive properties. The animal studies, however, tell us little about how serious a problem this may represent for human cannabis users, such information can only come from human studies, some of which is described in later chapters.

4

Medical Uses of Marijuana—
Fact or Fantasy?

Cannabis has been used as a medicine for thousands of years (Lewin, 1931; Walton, 1938; Robinson, 1996). The Chinese compendium of herbal medicines the *Pen ts'ao* first published around 2800 B.C. recommended cannabis for the treatment of constipation, gout, malaria, rheumatism, and menstrual problems. Chinese herbal medicine texts continued to recommend cannabis preparations for many centuries. Among other things its pain-relieving properties were exploited to relieve the pain of surgical operations.

Indian medicine has almost as long a history of using cannabis. The ancient medical text the *Athera Veda*, which dates from 2000–1400 B.C., mentions bhang (the Indian term for marijuana), and further reference is made to this in the writings of Panini (ca 300 B.C.).

> There appears to be no doubt that the cannabis plant was believed by the ancient Aryan settlers of India to possess sedative, cooling and febrifuge properties.
>
> (Chopra and Chopra, 1957)

In the ancient ayurvedic system of medicine cannabis played an important role in Hindu materia medica, and continues to be used by ayurvedic practitioners today. In various medieval ayurvedic texts cannabis leaves and resin are recommended as decongestant, astringent, soothing, and capable of stimulating appetite and promoting digestion. Cannabis was also used to induce sleep and as an anesthetic for surgical operations. It was also considered to have aphrodisiac properties and was recommended for this purpose.

In Arab medicine and in Muslim India frequent mention is also made of hashish (cannabis resin) and "benj" (marijuana). They were used to treat gonorrhea, diarrhea, asthma, and as an appetite stimulant and analgesic. In Indian folk medicine bhang (marijuana) and ganja (cannabis resin) were recommended as stimulants to improve staying power under conditions of severe exertion or fatigue. Poultices applied to wounds and sores were believed to promote healing, or when applied to areas of inflammation (e.g., piles) to act as an anodyne and sedative. Extracts of ganja were used to promote sleep and to treat painful neuralgias, migraine, and menstrual pain. Numerous concoctions containing extracts of cannabis together with various other herbal medicines continue to be used in rural Indian folk medicine today, with a variety of different medical indications including dyspepsia, diarrhea, sprue, dysentery, fever, renal colic, dysmenorrhea, cough, and asthma. Cannabis-based tonics with aphrodisiac claims are also popular. The use of cannabis-based medicines has declined rapidly in India in recent years, however. Their potency and effectiveness could never be guaranteed as the active ingredients (e.g., THC) vary unpredictably among different plant preparations, and they degrade on storage. At the same time reliable Western style medicines have become more generally available.

Cannabis or hemp was also popular in folk medicine in medieval Europe and was mentioned as a healing plant in herbals such as those by William Turner, Mattioli, and Dioscobas Taberaemontanus. One of the most famous herbals, written by Nicholas Culpepper (1616–1654) recommended that:

> . . . an emulsion of decoction of the seed. . . . eases the colic and always the troublesome humours in the bowels and stays bleeding at the mouth, nose and other places.

It was not until the middle of the nineteenth century, however, that cannabis-based medicines were taken up by mainstream Western medicine. This can be almost entirely attributed to the work of a young Irish doctor, William O'Shaughnessy, serving with the Bengal Medical Service of the East India Company. He had observed firsthand the many uses of cannabis in Indian medicine, and had conducted a series of animal experiments to characterize its effects and establish what doses could be tolerated. His experiments confirmed that cannabis was remarkably safe. Despite many escalations of dose it did not kill any mice, rats, or rabbits. O'Shaugnessy felt confident to go on to conduct studies in patients suffering from seizures, rheumatism, tetanus, and rabies. He found what appeared to be clear evidence that cannabis could relieve pain and act as a muscle relaxant and an anticonvulsant. The 30-year-old O'Shaugnessey reported his findings in a remarkable monograph, first published in Calcutta in 1839 and reprinted as a 40-page article in the Transactions of the Medical and Physical Society of Calcutta in 1842 (O'Shaugnessey,

1842). His report rapidly attracted interest from clinicians throughout Europe. As a result of his careful studies O'Shaugnessey felt able to recommend cannabis, particularly as an, "anticonvulsive remedy of the greatest value." He brought back a quantity of cannabis to England in 1842 and Peter Squire in Oxford Street, London was responsible for converting imported cannabis resin into a medicinal extract and distributing it to a large number of physicians, under O'Shaugnessey's directions.

O'Shaugnessy was a remarkable Victorian genius. John Moon gives a fascinating account of his career in an article published in the *New England Journal of Medicine* in 1967:

> O'Shaugnessy was born in Limerick in 1809 and like many Irishmen he went to Edinburgh for his medical education. There as a member of Professor Knox's anatomy class, he studied cadavers provided by the murderers Burke and Hare. Graduating at the age of twenty he soon demonstrated his talents in chemistry and toxicology and wrote on sulphocyanic acid, the presence of copper in organic matters and the detection of nitric acid and nitrate and potash. A year later he moved from Edinburgh to London, where he found himself a victim of prevailing unionism, unable to practice medicine within 7 miles of the city for want of a license from the Royal College of Physicians. . . . At this point O'Shaugnessy went to Newcastle-upon-Tyne where he began a series of experimental inquiries into the composition of the blood in cholera and on December 29, 1831 he wrote the following lines:
>
> > The blood drawn in the worst cases of the cholera is unchanged in its anatomical or globular structure.
> >
> > It has lost a large proportion of its water, 1000 parts of cholera serum having but the average of 860 parts of water.
> >
> > It has also lost a great proportion of its neutral saline ingredients.
> >
> > Of the free alkali contained in healthy serum, not a particle is present in some cholera cases, and barely a trace in others.
> >
> > Urea exists in the cases where suppression of urine has been a marked symptom.
> >
> > All the salts deficient in the blood, especially the carbonate of soda, are present in large quantities in the peculiar white dejected matters.

On January 7, 1832 he presented his data to the Central Board of Health in London and published them immediately thereafter in a brilliant monograph. This delightful book deserves a small place among medicine's classics as a demonstration of O'Shaugnessey's command of the literature, his clear and incisive logic, and his astonishing grasp of acid-base physiology. He related the functions of carbon dioxide, oxygen and the "colouring matter of the blood," and finally showed the essential elements of the chemical pathology of cholera. In his concluding remarks he stated, "I would not hesitate to inject some ounces of warm water into the veins. I would also without apprehension dissolve in that water the mild innocuous salts . . . which in cholera are deficient.

O'Shaugnessey himself never put these ideas to the test, but they were rapidly taken up by physicians and found to be effective. Cholera was a common and deadly infectious disease in nineteenth century cities that lacked modern sanitation systems. His ideas form the basis of the fluid replacement therapy, which to this day is the basis of treatment for the catastrophic loss of salts and water from the blood, which is a key feature of cholera and other diseases that induce severe diarrhea.

On moving to India in 1833 O'Shaugnessey began his studies of cannabis described above. In 1841, he published an important textbook of chemistry and was made professor of chemistry at the Medical College in Calcutta and 2 years later, at the remarkably young age of 34 he was elected a Fellow of the Royal Society in London.

His career then took another remarkable tack. In the late 1830s he experimented with telegraphy (for review see Bridge, 1998). At his own expense he constructed an experimental system using more than 30 miles of wire in the botanical gardens in Calcutta and devised practical methods for transmitting signals:

O'Shaugnessey tried for years to introduce the electric telegraph into India but could raise little enthusiasm until 1847, when he captured the interest of the great "proconsul" (or despot, depending on one's source of information), Lord Dalhousie. When he was promoted to surgeon in the Indian Medical Service medicine was no longer his principal concern. He obtained a commission from Dalhousie to lay down an experimental telegraph line between Calcutta and Kedjeree at the mouth of the Hooghly

River. The experiment proved immensely successful, and Dalhousie recommended the immediate construction of a network of lines linking Calcutta, Peshawar, Bombay, Agra and Madras. O'Shaugnessey was appointed director general of telegraphs and was deputed to England where at India House he and Sir Juland Danvers arranged for the needed men and materials. Within six months the line between Calcutta and Agra (800 miles) was fully operational. A year later (1855) 3500 miles of wire connected the major cities of the vast subcontinent, a notable feat for unskilled labour in terrain normally without roads and bridges. O'Shaugnessey returned to England where, in November 1856, Queen Victoria knighted him; but the greatest justification of his telegraph system came in the following year, when the Sepoy rebellion erupted. Most lines were promptly cut, but a single timely message to authorities in the Punjab summoned British troops under the command of Sir John Lawrence that soon recaptured Delhi to turn the tide in favour of the British Crown. Lawrence later wrote a poem entitled "The Telegraph Saved India."

(Moon, 1967)

By the time O'Shaugnessey retired to England in 1860, at the age of 51, there were 11,000 miles of telegraph lines in India and 150 offices in operation. Within 10 years telegraph links to London would be established. Despite all these achievements O'Shaugnessey was soon forgotten to posterity. In his biography J. A. Bridge (1998) wrote:

I think there is a reason for this. It is a peculiarity of the British culture that achievement in the Indian Empire should always stand at a discount, and O'Shaugnessey is no exception. Most British people have only the sketchiest of notions about the history of the British Indian Empire and how it came about. It seems that there never was great interest in the subject.

Following O'Shaugnessey's advocacy of cannabis and the availability of the medicinal extract it became popular for a while in British medical circles. Many doctors began to experiment with cannabis as a new form of treatment, and reports appeared in medical journals describing its application in a variety of conditions, including menstrual cramps, asthma, childbirth, quinsy, cough, insomnia, migraine headaches, and even in the treatment of withdrawal from opium. Cannabis extract and tincture appeared in the *British Pharmacopoeia* and were available for more than a hundred years:

> British Pharmaceutical Codex 1949:
> EXTRACTUM CANNABIS
> (Ext. Cannab.)
> Extract of Cannabis:
> Cannabis in coarse powder   1,000 g
> Alcohol (90%) .... a sufficent quantity
> Exhaust the cannabis by percolation with the alcohol and evaporate to the consistence of a soft extract. Store in well-closed containers, which prevent access of moisture.
> Dose: 16 to 60 milligrams
>
> TINCTURA CANNABIS
> (Tinct. Cannab.)
> Tincture of Cannabis:
> Extract of Cannabis   50 g
> Alcohol (90%)   to 1,000 ml
> Dissolve
>
> Weight per ml at 20°, 0.842 g to 0.852 g
> Alcohol content 83% to 87% v/v
> Dose 0.3 to 1 ml

In Britain, the eminent Victorian physician Dr. Russell Reynolds (Reynolds, 1890) recommended cannabis for sleeplessness, neuralgia, and dysmenorrhea (menstrual pains). It was also experimented with as a means of strengthening uterine contractions in childbirth and in treating opium withdrawal, an increasing problem for Victorian medicine as the uncontrolled consumption of opium increased, bringing with it problems of addiction. There was interest in the use of cannabis in the treatment of the insane, following reports by Dr. Jean Moreau in Paris of this possibility. But there was also concern that excessive use of cannabis could lead to insanity, a concern that persisted for many years — leading among other things to the Indian Hemp Drugs Commission's review of the use of cannabis in India (see Chapter 7).

Although Dr. Reynolds is said to have prescribed cannabis to Queen

Victoria to treat her menstrual pains, cannabis never really became popular in British medicine, and was used only infrequently. Difficulties in obtaining supplies and the inconsistent results obtained with different preparations of the drug made it hard to use. Because of the lack of any quality control to allow the preparation of standardized batches of the medicine, patients were likely to receive a dose that either had no effect or caused unwanted intoxication. Cannabis was never as reliable and widely used as opium — the mainstay of the Victorian medicine cabinet. Cannabis fell so far out of favor that it was the lack of any continuing medical use as much as any other factor which led to its removal from the *British Pharmacopoeia* by the middle of the twentieth century.

In America, cannabis was already known even before O'Shaugnessey made it popular in Europe. It was first introduced into homeopathic medicine, as described in the *New Homeopathic Pharmacopoeia and Posology or the Preparation of Homeopathic Medicines*:

> To make the homeopathic preparation of hemp we take the flowering tops of male and female plants and express the juice, and make the tincture with equal parts of alcohol; other advise only to use the flowering tops of female plants, because these best exhale, during their flowering, a strong and intoxicating odour, whilst the male plants are completely inodorous.
>
> (Jahr, 1842)

Cannabis came to the notice of psychiatrists also, who experimented with its use in treating the mentally ill. By 1854 the United States Dispensatory began to list cannabis among the nation's medicinals, and gave the following remarkably accurate description of its properties:

> Medical Properties: Extract of hemp is a powerful narcotic, causing exhilaration, intoxication, delirious hallucinations, and, in its subsequent action drowsiness and stupor, with little effect upon the circulation. It is asserted also to act as a decided aphrodisiac, to increase the appetite, and occasionally to induce the cataleptic state. In morbid states of the system, it has been found to produce sleep, to allay spasm, to compose nervous inquietude, and to relieve pain. In these respects it resembles opium in its operation; but it differs from that narcotic in not diminishing the appetite, checking the secretions, or constipating the bowels. It is much less certain

in its effects; but may sometimes be preferably employed, when opium is contraindicated by its nauseating or constipating effects, or its disposition to cause headache, and to check the bronchial secretion. The complaints to which it has been specially recommended are neuralgia, gout, tetanus, hydrophobia, epidemic cholera, convulsions, chorea, hysteria, mental depression, insanity, and uterine hemorrhage. Dr. Alexander Christison, of Edinburgh, has found it to have the property of hastening and increasing the contractions of the uterus in delivery, and has employed it with advantage for this purpose. It acts very quickly, and without anesthetic effect. It appears, however, to exert this influence only in a certain proportion of cases.

(Wood and Bache, 1854)

Although cannabis continued to attract the interest of psychiatrists, cannabis did not become widely popular with American doctors. During the Civil War it was used to treat diarrhea and dysentery among the soldiers, but as a medicine cannabis had too many shortcomings. As British doctors had found, the potency of commercial preparations varied from pharmacist to pharmacist as there was no means of standardizing the preparations for their content of the active drug. What proved to be an effective dose when using material from one supplier would either have no noticeable effects or would produce unpleasant intoxication when the same amount was obtained from a different supplier. In addition, the drug was not water soluble and so unlike morphine, which was then becoming available, cannabis could not be given by injection. The hypodermic syringe was invented in the late nineteenth Century and was immediately popular with doctors and patients for administering instant remedies. There is a certain mystique associated with ritual of an injection — even today many Japanese patients prefer their medicines to be administered in this way. Cannabis had to be given by mouth and took some time to take effect. The doctor might have to remain with his patient for more than an hour after giving the drug, in order to make sure not only that it was having the desired effect but also that the dosage had not been too high.

A succinct and perceptive summary of the rise and fall of cannabis in nineteenth century medicine is given by Walton (1938, p.152):

The popularity of the hemp drugs can be attributed partly to the fact that they were introduced before the synthetic hypnotics and analgesics. Chloral hydrate was not introduced until 1869 and was followed in the next 30 years by paraldehyde, sulfonal and the barbitals. Antipyrine and acetanilide, the first of their particular group of analgesics (aspirin-like drugs), were introduced about 1884 (aspirin, not until 1899). For general sedative and analgesic purposes, the only drugs commonly used at this time were the morphine derivatives and their disadvantages were very well known. In fact, the most attractive feature of the hemp narcotics was probably the fact that they did not exhibit certain of the notorious disadvantages of the opiates. The hemp narcotics do not constipate at all, they more often increase rather than decrease appetite, they do not particularly depress the respiratory center even in large doses, they rarely or never cause pruritis or cutaneous eruption and, most importantly, the liability of developing addiction is very much less than with the opiates.

These features were responsible for the rapid rise in popularity of the drug. Several features can be recognised as contributing to the gradual decline of popularity. Cannabis does not usually produce analgesia or relax spastic conditions without producing cortical effects and, in fact, these cortical effects usually predominate. The actual degree of analgesia produced is much less than with the opiates. Most important, the effects are irregular due to marked variations in individual susceptibility and probably also to variable absorption of the gummy resin.

Pharmaceutical companies, nevertheless, tried to make use of cannabis as a medicine and it was included in dozens of proprietary medicines, which were available over the counter in the nineteenth century and the early years of the twentieth century. These included the stomach remedy *Chlorodyne* (which also contained morphine; Squibb Co.), *Corn Collodium* (Squibb Co.); *Dr. Brown's Sedative Tablets*, and *One Day Cough Cure* (Eli Lilly Co). The company Grimault and Sons marketed cannabis cigarettes as a remedy for asthma. By 1937, when cannabis was removed from medical use in the United States some 28 different medicines contained it as an ingredient — many of them with no indication of its presence.

The Modern Revival of Interest in Cannabis-Based Medicines

During most of the twentieth century there has been little interest in the use of cannabis in Western medicine, and such use has been legally prohibited since 1937 in the United States and since the 1970s in Britain and most of Europe. As cannabis became an increasingly popular recreational drug during the 1960s and 1970s, however, more and more people were exposed to it and during the 1980s and 1990s there has been an increasing interest in medical applications. As the British Medical Association in their influential report on the therapeutic uses of cannabis published in 1997 puts it:

> . . . many normally law abiding citizens—probably many thousands in the developed world use cannabis illegally for therapy.

The groups most commonly involved in such illegal self-medication are those suffering from chronic pain conditions that are unresponsive to other pain-relieving drugs, patients with spinal injuries or other spastic conditions who frequently experience painful muscle spasms, patients with AIDS and sufferers from multiple sclerosis (MS). The British MS Society conducted a survey of their members that showed that 1% took medicinal cannabis, but they believe that the real figure is higher—perhaps as many as 4%. With some 80,000 MS patients in Britain, this would represent more than 3000 self-medicating illegally with cannabis. Most of the patients who are taking cannabis in Europe and in the United States use smoked marijuana—in contrast to the earlier use of the drug in Western medicine, which was invariably taken by mouth and not smoked.

With many centuries of experience of cannabis as a safe medicine, and with thousands of patients in Western countries convinced of the benefits of the drug, why is there any problem? Why do Western governments not agree to make it legally available for doctors to prescribe for their patients? Governments have clearly stated political reasons for withholding such consent, as they do not wish to give the wrong message to young people. If teenagers see governments approving cannabis as a drug that is safe to use in patients would this not encourage even greater illicit recreational use? But are there any scientific reasons to withhold a safe and effective medicine from patients?

The problem is that although cannabis has been used in human medicine for some 4000 years we do not have rigorous scientific evidence either for its safety or its effectiveness except in a very few isolated instances. The fact of the matter is that many folk medicines and herbal remedies have no real beneficial medical effects, they are used because of tradition and folklore, and in many cases if patients show some improvement in their symptoms this says more for the power of suggestion than the efficacy of the medicines. This is seen par excellence with homeopathic medicines. These consist of a variety of natural ingredients used in very dilute form. Homeopathic medicine is based on two principles, the "Law of Similars," which states that the treatment of illness should involve the administration of substances that themselves can cause the symptoms of the illness. This does not happen, however, because the second "Law of Infinitesimals" states that the more dilute the medicine is the more effective it will be. It is not uncommon for homeopathic preparations to have been diluted serially (10-fold at each step) as many as 30 times. In effect this means that the bottle of homeopathic medicine may contain only a few molecules (or maybe even none) of the original natural medicine. Since many of the ingredients used are highly toxic (e.g., bee venom, tarantula spider venom, silver nitrate, arsenic oxide) this may be just as well! To the pharmacologist who knows that drugs invariably exhibit dose–response curves in which the biological response increases with increasing drug concentration, not with dilution, this is meaningless mumbo jumbo. It is clear that most homeopathic medicines can have no biological effect, yet many patients are convinced that they derive benefit, and are willing to believe the complex pseudoscientific explanations of how particular homeopathic remedies are formulated for particular illnesses. There are training schools for practitioners of homeopathic medicine, there are homeopathic pharmacopeias, and in some European countries homeopathic remedies account for more than 10% of the total medicines market. In the United States homeopathic medicines are gaining in popularity and have special exemption from the Food, Drug and Cosmetics Act of 1938 which normally restricts the sale of medicines to those approved by the government's Food and Drug Administration (FDA). The FDA issued guidelines in 1988 concerning the sale of homeopathic medicines, recommending that they must not claim to cure such life-threatening conditions as cancer or AIDS. Some opponents would like to see homeopathic medicines banned, but the FDA declined to grasp this nettle on the grounds that these medicines are considered safe (as they contain essentially no active ingredients) and their efficacy has not been proved or disproved. Many famous people, including Tina Turner, Lauren Hutton, Larry King, and Queen Elizabeth II, have publicly affirmed their belief in the value of homeopathic medicine. Some private medical insurance companies [e.g., British United Providence Association (BUPA)] are willing to reimburse their members if they consult a homeopath. The Spring 1999 issue of BUPA's magazine *Upbeat* contained the following case history, lending respectability to this branch of alternative medicine:

> Sarah, 16, suffered from eczema since birth — along with hay fever and a house-dust allergy.
>
> "I had eczema all over my body and face. I used steroid creams and inhalers, but I was worried about the effects of the creams on my skin. I was also fed up with the continual itching, sneezing and wheezing. I saw a homeopath who gave me a sulphur remedy. Although it made my eczema worse to begin with, I was amazed at the improvement in a matter of weeks. For the past five months I have been using a steroid cream once a week instead of every day, and during the summer I hardly used my inhaler at all. I need to continue seeing the homeopath once every three months for about a year, but I'm really pleased with the results so far."

Homeopathic medicine has no scientific rationale. Nevertheless, if patients believe that a treatment will benefit them, this belief and the optimism with which it imbues them can have powerful effects on the course of an illness.

The effects of homeopathic medicines most likely involve this well-documented placebo effect. If people are given a tablet or capsule identical to that containing a genuine medicine, but which contains no active ingredients other than sugar or some other inert powder, they will often report that they feel better. This even extends to the treatment of severe pain, where patients receiving placebo may report pain relief. Some years ago Jon Levine and Howard Fields, researchers at the University of California in San Francisco, conducted some ingenious experiments that throw some light on the mysterious placebo effect. They studied groups of students who had attended the student dental clinic for the surgical removal of wisdom teeth. The students were told that they would receive either an inactive placebo or the powerful painkiller morphine. Two hours after recovering from the anesthetic, the subjects who had volunteered to take part in this study received an intravenous injection of either morphine or a saline placebo. To ensure that the investigator did not inadvertently reveal whether the subjects were receiving morphine or placebo, the study was *blinded*, i.e., neither the subjects nor the physician knew which subjects were receiving the active drug. This information was coded and held by someone not involved directly in the experiment and the code was only broken when the experiment was complete. After a dental operation, most people experience pain that increases gradually over a period of several hours. Those subjects who received morphine reported that their pain was either stable or decreased. The subjects who received the placebo saline injection fell into two groups. About two thirds of them showed no response and their pain increased gradually over the course of the study, but about one third of the placebo group were classified as responders since they reported pain relief that was equivalent to subjects who had received a moderate dose of morphine (Levine et al., 1979) (Fig. 4.1). In another study it was found that the drug naloxone, which acts a potent antagonist of the actions of morphine and related opiates at the opiate receptors in brain and spinal cord, could prevent the placebo response in placebo responders, but it had no effect in placebo nonresponders. How could naloxone block the effect of a drug that the placebo group had not received? The answer seems to be that the mere expectation of pain relief from an injection that might contain morphine was by itself sufficient in some people to activate the

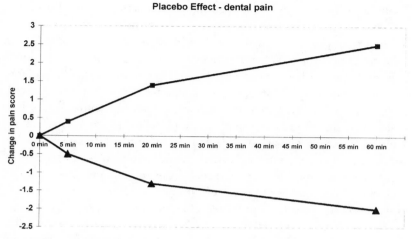

Figure 4.1. Placebo effect. Subjects received an injection of either morphine or saline (placebo) in a blinded manner 2 hours after dental surgery, and were asked to rate their pain scores on an arbitrary scale for the subsequent hour. Data are shown only for the group receiving placebo. Of 107 patients in this group 42 (39%) were rated as placebo responders. Whereas nonresponders experienced an increasing level of pain (filled squares), the placebo responders either reported some degree of pain relief (filled triangles) or their pain remained unchanged (data not shown). Redrawn from Levine et al. (1979).

body's own natural opiate system, causing the release of the morphine-like chemicals known as enkephalins in the brain and spinal cord. This produced pain relief, but the enkephalins were ineffective in the presence of naloxone, which blocks the receptors through which they act.

The San Francisco study gives us some hint of how some genuine placebo effects may be explained. It also illustrates some of the principles underlying modern clinical trials. The introduction of new medicines for human use requires that they fulfill internationally agreed criteria for safety and effectiveness laid down by the various government regulatory agencies responsible for the approval of new medicines. The effectiveness of the medicine in treating a particular illness has to be established in controlled clinical trials. Controlled means comparing the test drug with an inactive placebo prepared in such a manner that it cannot be distinguished from the active medicine. In a double-blind, randomized, placebo-controlled trial neither the patient nor the doctor nor the nurse knows whether active drug or placebo is given to any particular patient. This information is held in coded form by a person not actively involved in the conduct of the trial and is not made available until the trial has ended. Patients are randomly allocated to placebo and test-drug groups to avoid any possible bias in the selection of those who are to receive the active drug. The outcome of the trial should involve objective measurements wherever possible, using predetermined outcome measures or endpoints. The success or failure of the trial is measured by criteria established in a written trial protocol before the start of the trial. Because individual patients will inevitably vary in their response to drug or placebo, the trial should include a sufficiently large number of subjects to provide statistically significant differences in outcome measures between the placebo and the drug-treated groups.

The modern era of clinical trials is often thought to have started in 1948, with the publication of the United Kingdom Medical Research Council trial of the drug streptomycin in the treatment of tuberculosis. To avoid bias, the decision on whether a patient would receive streptomycin (a new antibiotic at that time) and bed rest, or bed rest alone, was made by opening the next in a series of envelopes. These had been prepared earlier in random order. Neither doctor nor patient could know which treatment was given as the envelopes were filled and opened by an independent person. The results showed streptomycin to be highly effective.

There are a number of variants on clinical trial design. For example, it is not always necessary to use separate groups of patients to assess test-drug or placebo responses. In the so-called "crossover" trial design the same patients receive placebo and test drug at different stages during the trial and are crossed over from one to the other after a "wash-out" period (to ensure removal of any active drug from the body). The test drug or placebo is given to different patients in random order, so that the trial remains double blind.

These principles of clinical trial design, although they may appear to be simply common sense, are relatively new. It is only in the past 40 years that the concept of the controlled clinical trial has become generally accepted. It can be applied not only to the testing of new medicines, but also to the effectiveness of any new medical procedure—although there continues to be resistance to the concept of evidence-based medicine among some members of the medical profession.

The reasons for insisting on these elaborate scientifically controlled trials was the growing realization that the expectations of both doctor and patient can influence the outcome of a clinical trial, even though neither may be consciously aware of this. The importance of the placebo effect means that this has to be built into the design of the trials. Not all human illnesses will show the same degree of susceptibility to the placebo effect, such treatment is most likely to affect the outcome of conditions in which there are strong psychological or psychosomatic components, and less likely to influence the outcome of infectious diseases or cancer. Placebo effects are particularly prominent in the treatment of such psychiatric conditions as anxiety and depression, and are often seen in illnesses in which the patient has failed to gain any benefit from existing conventional medicines. Such patients are often desperately seeking new treatments, which they want to work.

There is a real possibility that some of the medical benefits claimed by patients who are self-medicating with cannabis could lie in that category. The patients are usually those for whom conventional medicine has failed and they are turning to alternative medicine for relief from their symptoms. Cannabis has the added attraction to many of being a natural and herbal remedy, embedded in centuries of folklore and folk medicine. At the moment hardly any of the medical indications for which herbal cannabis is illegally used can be substantiated by data from scientifically controlled clinical trials. The thousands of patients who are currently self-medicating rely almost entirely on word of mouth anecdotal evidence and their own personal experiences of the drug. Anecdotal evidence, however, is not reliable and cannot be used to persuade regulatory agencies to approve cannabis as a medicine. To the nonscientist this is hard to understand. The often moving personal accounts of individuals who report the benefits they have derived from herbal cannabis are so compelling — what more is needed?

The Synthetic Cannabinoids

In the sound and fury of the debate about the medical use of herbal cannabis, with strongly held positions on both sides of the argument, it is often forgotten that two cannabis-based medicines are already available by prescription to patients on both sides of the Atlantic. These are the synthetic cannabinoids dronabinol (Marinol®) and nabilone (Cesamet®). Although they have not proven very popular, the medical use of these compounds, unlike herbal cannabis is backed up by a substantial body of scientific evidence from clinical trials, and the compounds have satisfied the strict requirements of the United States Food and Drug Administration for approval as human medicines.

Dronabinol (Marinol®)

Dronabinol is the generic name given to Δ^9-THC (Fig. 2.1). It is marketed as the medical product known as Marinol®. Drugs are given an official generic name — which is used when describing the compound in the scientific literature, and the company that markets the drug usually gives it a separate trade name. The same drug may be marketed by more than one company under different trade names, but each compound can only possess one generic name.

One of the problems in using dronabinol as a medicine is that the pure compound is a viscous pale yellow resin, which is almost completely insoluble in water. This makes it impossible to prepare it as a simple tablet and it cannot be dissolved for administration as an intravenous injection. Marinol® is, therefore, prepared by dissolving dronabinol in a small quantity of harmless sesame oil in a soft gelatine capsule (containing 2.5, 5, or 10 mg dronabinol). These capsules are easily swallowed, and once in the stomach the gelatine dissolves releasing the drug. The oil forms an emulsion of small drug-containing droplets from which the drug is absorbed during passage through the gut. Absorption is almost complete (90%–95%) but because much of the active drug is metabolized during passage of the blood from the gut via the liver, only 10%–20% of the administered dose reaches the general circulation. Effects begin after 30 minutes to 1 hour and reach a peak at 2–4 hours, with duration of action of 4–6 hours, although the appetite stimulating effects of the drug may persist for up to 24 hours. Considerable quantities of the psychoactive metabolite 11-hydroxy-THC (see Chapter 2) are formed in the liver, and this metabolite is present in blood at approximately the same level as the parent drug, with a similar duration of action.

Two medical indications have been approved for dronabinol. The first of these is its use to counteract the nausea and vomiting frequently associated with cancer chemotherapy, the other is as an appetite stimulant to counteract the AIDS wasting syndrome, as described below (for review see Plasse et al., 1991). The annual sale of Marinol® in the United States currently is estimated at $20 million. About 80% of prescriptions are for HIV/AIDS patients, 10% for cancer chemotherapy, and 10% for a range of other purposes.

The possibility that medical supplies of dronabinol might be diverted to illicit use has been a concern, but there is very little evidence that this has happened. Dronabinol has little value as a street drug. The onset of action is slow and gradual and its effects are unappealing to regular marijuana smokers; it has a very low abuse potential. Because of the low dependence liability, the Drug Enforcement Agency (DEA) in November 1998 announced a proposal to reschedule Marinol® to the less restrictive Schedule III.

Nabilone (Cesamet®)

During the 1970s a number of pharmaceutical companies carried out research on synthetic analogues of THC to see whether it might be possible to dissociate the desired medical effects from the psychotropic actions. On the whole this quest proved disappointing, and in retrospect this may have been inevitable since we now believe that both the desired effects and the intoxicant actions of THC result from activation of the same CB1 receptors in the central nervous system. Only one company persisted with this research long enough to produce a marketed product — nabilone (Cesamet®), (Fig. 2.2). Nabilone is a potent analogue of THC, which scientists at the Eli Lilly Company believed might have an improved separation of the desired therapeutic effects from psychotropic actions. Unlike dronabinol, nabilone is a stable crystalline solid, and for human use the drug is prepared in solid form in capsules containing 1 mg of nabilone that are taken by mouth, and the dose is usually 1 or 2 mg twice a day.

Preliminary clinical studies in the treatment of anxiety gave promising results, but the company decided to focus on the treatment of nausea and vomiting in patients undergoing cancer chemotherapy as the primary target. They carried out the most complete series of controlled clinical trials so far undertaken on any cannabinoid, as described below (for review see Lemberger, 1985)

What Are the Medical Targets for Cannabis?

Nausea and Vomiting Associated with Cancer Chemotherapy

Ironically the condition for which there is the strongest scientific evidence for beneficial effects of cannabis-based medicines is also now no longer an area of pressing medical need since new and even more powerful antisickness drugs have become available recently. When the cannabinoids dronabinol and nabilone were first being tested, in the 1970s and early 1980s, however, matters were different. The treatment of cancer with more and more powerful drugs to suppress the growth of tumor cells advanced considerably during the 1960s and 1970s. Although the newer chemotherapy drugs were increasingly effective as anticancer agents they brought with them severe side effects. As the 1997 *British Medical Association Report* puts it:

> One of the most distressing symptoms in medicine is the prolonged nausea and vomiting which regularly accompanies treatment with many anti-cancer

agents. This can be so severe that patients come to dread their treatment; some find the side-effects of the drugs worse than the disease they are designed to treat; others find the symptoms so intolerable that they decline further therapy despite the presence of malignant disease.

One of the most effective anticancer drugs is the platinum-containing compound cisplatin, but unfortunately it is also very powerful in causing nausea and vomiting. Cancer patients receiving this drug almost invariably experience nausea and vomiting, with an average of six bouts of vomiting during the first 24 hours unless they are protected by antiemetic medicines.

The results of properly controlled clinical trials conducted in the 1970s and 1980s indicated that the two cannabinoid drugs dronabinol and nabilone appeared to offer a potentially important advance over the relatively ineffective antisickness medicines available in the early 1980s. The most widely used drugs then were chlorpromazine, prochlorperazine, haloperidol, metoclopramide and domperidone — all of which act as antagonists of the chemical messenger dopamine. In the various clinical trials in which dronabinol was compared with placebo or with another antisickness agent, prochlorperazine, a total of 454 patients suffering from various forms of cancer received the drug. Dronabinol doses ranged from 2.5 mg/day to 40 mg/day, given as equally divided doses every 4–6 hours. Approximately two-thirds of the patients experienced complete or partial relief from nausea and vomiting, but at the higher doses disturbing psychotropic effects became apparent in many patients. The optimum dose regime for most patients seems to be 5 mg three or four times a day (for review see *British Medical Association Report,* 1997). The use of dronabinol as an antisickness agent is supported by the results of animal experiments that show its effectiveness in various animal models, although the precise site of action in the brain remains unknown.

With nabilone some 20 separate clinical trials involving more than 500 patients were reported, many with a double-blind crossover design to allow the direct comparison of nabilone with prochlorperazine or other antiemetic medicines in the same patients. Nabilone proved to be as effective, or more so, as prochlorperazine and it successfully treated the symptoms of nausea and vomiting in 50%–70% of patients. Central nervous system side effects of drowsiness, light-headedness, and dizziness were seen in more than half of the patients, but these were not considered serious, and only a small proportion of patients (about 15%) experienced a high (Lemberger, 1985). The company believed that the drug could be used successfully as an antiemetic without causing intoxication, and they were successful in gaining approval from the FDA to market this product. The United States Drug Enforcement Agency, however, concluded that nabilone was still too much like cannabis, and they gave it a restrictive Schedule II classification, i.e., it is considered a potentially dangerous drug of addiction although it does have some medical usefulness. The Schedule II classification was disappointing to Eli Lilly, as it placed onerous requirements on the company and any physicians using the compound to keep it securely and to record its every movement. The company lost interest in further research in the area and did not place any major marketing effort behind nabilone, which has had little popularity.

These data, from properly controlled clinical trials are important because they show that it is possible to obtain such results with cannabis-based medicines, but dronabinol and nabilone have not proven popular in clinical use. The effective dose of either cannabinoid as an antiemetic is too close to the dose that causes sedation or intoxication, and this limits the amount of drug that can be given. Patients who have not had any previous exposure to cannabis generally find the psychotropic effects of the drug unpleasant and disturbing. The value of dronabinol and nabilone in the treatment of nausea and vomiting associated with cancer therapy has also been eclipsed by the development, during the 1980s, of new and more powerful antisickness drugs. There is now a series of such drugs, which act by blocking one of the receptors for the chemical messenger serotonin. The 5-HT$_3$ receptor that is targeted by these compounds plays a key role in the neural circuits in the nervous system involved in the vomiting reflex. The first such drug ondansetron (Zofran®) proved highly successful and was followed by others: e.g., granisetron (Kytril®) and tropisetron (Navoban®). The serotonin antagonists have several advantages over dronabinol. They do not suffer from the psychotropic side effects that limit the usefulness of the cannabinoid, and they are able to control nausea and vomiting in a larger proportion of patients. In addition, unlike the water-insoluble cannabinoid, the serotonin antagonists can be dissolved easily for intravenous injection. They are commonly used as an initial intravenous injection at the time of the cancer chemotherapy or radiation therapy, followed by oral tablets for the next few days. The serotonin antagonists, together with the steroid dexamethasone with which they are often combined, satisfactorily control nausea and vomiting in 80%–90% of cancer patients. The introduction of these new drugs has radically improved cancer therapy and they have become widely used. A survey of more than 1000 cancer specialists in the United States published in 1997 reported that more than 98% of them had prescribed serotonin antagonists more than five times between 1992 and 1994, whereas only 6% had prescribed dronabinol during this period (Schwartz et al., 1997). There is no longer much demand for dronabinol with its unpredictable side effects and variable absorption, although there may still be a niche for new medicines to treat the small proportion of cancer patients who do not respond well to the serotonin antagonists.

Limited clinical trials with another potent synthetic cannabinoid, levonantrodol showed that this compound was also effective as an antiemetic, but its CNS side effects limited its usefulness. Some excitement was generated by the results of clinical trials in Israel of the close relation of dronabinol, Δ^8-THC. In a group of eight children receiving cancer chemotherapy all responded well to the oral administration of this cannabinoid, with minimal CNS side effects (Abrahamov et al., 1995). This result has not been followed up, however, and there seems no reason to believe that Δ^8-THC is less likely to cause CNS side effects than dronabinol.

A few clinical trials have attempted to assess the effectiveness of smoked marijuana in controlling the symptoms of nausea and vomiting in patients undergoing cancer chemotherapy. Such trials are naturally difficult to blind, although some studies have used placebo marijuana cigarettes, using herbal cannabis from which THC had been extracted beforehand. The Research Triangle Institute in North Carolina, was commissioned by the United States government in the 1970s to produce standardized marijuana cigarettes, of consistent THC content (using herbal cannabis grown by the University of Mississippi), and placebo cigarettes for use in clinical research studies. Experienced marijuana users, however, have little difficulty in distinguishing the THC-containing smoked material from the placebo, making it hard to undertake a properly blinded trial. Partly because of these difficulties, very few controlled clinical trials have ever been described (see National Institutes of Health [NIH] Report on the Medical Uses of Marijuana, 1997 and American Medical Association Report on Medical Marijuana, 1997). In one study, smoked marijuana and orally administered dronabinol were compared in a random-order crossover design trial in 20 cancer patients. The treatments were effective in only 25% of the patients. When asked for their preference for the oral versus smoked drug, 45% had none, 35% preferred dronabinol, and only 20% preferred smoked marijuana (Levitt et al., 1984). In a larger open trial, smoked marijuana was tested in 74 cancer patients who had not responded well to other medicines. Eighteen patients dropped out of the trial because they found marijuana smoke too harsh and preferred oral dronabinol. Of the remaining 56 patients, 18 (34%) rated smoked marijuana very effective and 26 (44%) rated it moderately effective. Twelve (22%) reported no benefit. Sedation occurred in 88% of the patients, dry mouth in 77%, and dizziness in 39%, only 13% were free of adverse side effects (Vinciguerra et al., 1988). In another trial, smoked marijuana was offered to cancer patients who continued to experience nausea and vomiting after having taken oral dronabinol. Smoked marijuana (every 3–4 hours for several days) had a significant antiemetic effect, and the plasma levels of THC were higher after smoking than after oral dronabinol. But this study was performed in young patients (median age = 24 years) most of whom had prior experience with smoked marijuana. In contrast, another trial reported negative results with smoked marijuana in an older group of patients (median age = 41 years) who were inexperienced in using the smoked drug. Probably because of their inexperience, the plasma levels of THC that were achieved were low.

During the late 1970s and early 1980s a number of state departments of health in the United States conducted open-label studies of smoked marijuana, using protocols approved by the FDA. Such studies were carried out in California, Georgia, Michigan, New Mexico, New York, and Tennessee in a total of 698 cancer patients, most of whom had not responded well to other antiemetic medicines. Unfortunately these large studies were not well controlled, there was no attempt to use placebo, and the outcome was based not on objective measurements but on patient and/or physician's ratings. Nevertheless, smoked marijuana was reported to be comparable or more effective than orally administered dronabinol and more effective than prochlorperazine or other antiemetics available at that time in reducing nausea and vomiting. When given a choice, many patients preferred smoked marijuana to oral dronabinol. The most common side effect with smoked marijuana was sedation, whereas oral dronabinol tended to cause more unpleasant CNS side effects (dysphoria and dizziness).

It seems unlikely that smoked marijuana has an important role to play in controlling the symptoms of nausea and vomiting associated with cancer chemotherapy. Although smoking marijuana may offer a more precise control of plasma levels of THC in the hands of experienced smokers, this is not a method of drug delivery likely to be effective or acceptable in elderly drug-naïve patients. It is also hard to conceive of such a practice in modern smoke-free hospitals, although some patients will continue to prefer to self-medicate with smoked marijuana — claiming that it allows them to titrate the right dose to control their symptoms. There is no doubt that some gain a genuine benefit:

Harris Taft was receiving chemotherapy for Hodgkin's disease. His wife, Mona Taft, gives the account:

"One day in 1977, when we arrived at the treatment room where Harris was to receive the injection, he bolted and ran down the corridor. I found him a bit later, wandering the halls. He told me he couldn't take any more chemotherapy. He was at wit's end, exhausted by the disease, terrified by the effects of the drugs that were supposed to prolong his life. I have never before or since seen a man so genuinely and deeply frightened. Harris had come to fear the treatments more than the cancer and, he admitted, more than death itself. He told me he would choose dying over further chemotherapy."

Later Harris tried smoking cannabis before chemotherapy; it completely controlled the vomiting.

"It is impossible for me to adequately describe what a profound difference marihuana made. Before using marihuana, Harris felt ill all the time, could not eat, could not even stand the smell of food cooking. Afterward, he remained active, ate regular meals, and could be himself. His mood, his manner, and his overall outlook were transformed. And of course, marihuana prolonged his life by allowing him to continue chemotherapy. In two years of smoking it, he never had an adverse or untoward reaction. Marihuana was the least dangerous drug my husband received during the nine years he was treated for cancer."

(Grinspoon and Bakalar, 1993. This and other excerpts are reprinted with permission of the publishers and copyright holders, Yale University Press)

The case for orally administered cannabinoids in the treatment of patients undergoing cancer chemotherapy seems stronger. However, the clinical trial data show that one third of cancer patients failed to respond to the cannabinoid drugs and many found the psychotropic side effects intolerable. The inability to administer the water insoluble cannabinoids intravenously is also an important limitation. The advent of the serotonin antagonists, as described earlier, essentially put an end to any enthusiasm that the medical profession had for cannabinoids. Nevertheless, there remains a need for new antiemetic agents to treat the 10%–20% of cancer patients whose symptoms are not controlled well by the serotonin antagonists. These drugs are also only partially effective in controlling the delayed nausea and vomiting that persist for several days after cancer chemotherapy. It is possible that adding a low (nonpsychotropic) dose of cannabinoid to the dose regime might make it possible to treat some of the patients who fail to respond to serotonin antagonists alone. Some clinical trial data already point to an enhanced effectiveness when a cannabinoid was combined with prochlorperazine. But there has been no research comparing cannabinoids with serotonin antagonists or testing combinations of these drugs. Meanwhile entirely new approaches to the treatment of nausea and vomiting are emerging in the shape of a class of drugs that targets the peptide messenger substance P, which like serotonin is involved in the neural circuits that control the vomiting reflex. Preliminary clinical trial results with one of the new substance P antagonist drugs seem very promising and these agents could prove more effective than the serotonin antagonists in controlling both the early and the delayed stages of the sickness that accompany cancer chemotherapy.

AIDS Wasting Syndrome

Loss of appetite and a progressive involuntary weight loss of about 10% of body weight are seen in AIDS wasting syndrome, a characteristic feature of the disease. The onset of bouts of wasting syndrome, which last for a month or more, is one of the defining events in the transition from HIV positive to AIDS. The wasting is accompanied by chronic diarrhea, weakness, and fever. The advent of the newer and more powerful treatments for AIDS may make the wasting syndrome less common in the future, but it currently remains a distressing feature of the disease. Although the precise physiological mechanisms underlying the wasting syndrome are not well understood, the loss of weight seems to be primarily due to reduced energy intake.

There has been considerable interest in the use of both smoked marijuana and oral dronabinol as appetite stimulants for AIDS patients. An increased appetite, particularly for sweet foods, occurring about 3 hours after smoking marijuana is well known anecdotally as a feature of marijuana intoxication. Placebo controlled studies with smoked marijuana in normal healthy volunteers have confirmed that this is a genuine phenomenon. The mechanism involved is not known, but seems to involve a combination of enhanced hunger and an increased sensory attractiveness of the foods. Repeated dosing of healthy volunteers stimulated appetite and caused a measurable increase in caloric intake.

Experiments in animals, however, have failed to show any consistent effects of THC on appetite or body weight.

The second approved indication for dronabinol is as an appetite stimulant to treat the loss of appetite and weight loss associated with AIDS. After a series of small-scale clinical trials gave promising results, a larger placebo controlled clinical trial was conducted in 139 such patients (Beal et al., 1995). As compared to placebo, dronabinol treatment resulted in a statistically significant improvement in appetite after 4 to 6 weeks of treatment, and this effect persisted in those patients who continue receiving dronabinol after the end of the formal trial. There were trends toward increases in body weight and a decrease in nausea. The dose of Marinol® that appears to be optimum is 5 mg per day, administered as two doses of 2.5 mg, one given 1 hour before lunch, and one given 1 hour before supper. Other clinical trial data suggest that dronabinol may also benefit AIDS patients suffering nausea and loss of appetite as a consequence of treatment with antiviral drugs. Other clinical studies have indicated that dronabinol causes a significant stimulation of appetite in cancer patients, who also commonly suffer loss of appetite and an accompanying body weight loss. In both cancer patients and in AIDS patients suffering from wasting syndrome it is difficult to know whether the beneficial effects of dronabinol are not due in part at least to its ability to treat the symptoms of nausea that often accompany these syndromes. Dronabinol does not seem likely to be of any benefit to patients suffering from anorexia nervosa. The results of a controlled double blind crossover design trial in 11 such patients in which dronabinol was compared with diazepam (Valium®) found no benefit with either drug, and dronabinol caused unpleasant psychotropic reactions in three of the patients.

As in the treatment of nausea and vomiting, the principal adverse side effect in the use of dronabinol as an appetite stimulant has been the intensity of the accompanying CNS side effects. While careful choice of the optimum dose and its timing relative to meals can manage these in some patients, the delayed onset of action of the orally administered drug and its long duration of action are negative features. For this reason it is not surprising that many AIDS patients have turned to self-medication with smoked marijuana. Unfortunately there are few controlled studies that have attempted to document objectively the benefits of such treatment, although it is believed to be widespread. There is only anecdotal evidence:

> Ron Mason was diagnosed with hepatitis B in 1983 and later with HIV. Following his diagnosis of hepatitis B he found cannabis helpful in controlling his symptoms of nausea and vomiting:
>
> > "Although I lacked appetite, the doctor told me that I had to eat. Since I had a liver disease, I naturally gave up drinking (I had never drunk much anyway), and now began to smoke more marihuana. I noticed that my appetite increased dramatically after smoking. I began to smoke daily and gained weight rapidly. Two years later I had not produced antibodies and was officially designated a hepatitis carrier.
> >
> > . . . In April 1984 I was referred by my doctors at a gay VD clinic to what was later to become known as the AIDS clinic in Chicago. I saw doctors there for seven years and gained 40 pounds, achieving normal weight. The doctors knew that I smoked marihuana and did not forbid it, although they urged moderation. I cannot tolerate AZT because of anaemia. All the other antiviral drugs are damaging to my hepatitis-infected liver.
> >
> > Three years ago one of my doctors told me that I am one of a handful of people who have been going to the clinic for several years and are not dead or gravely ill; the doctors don't know why. I attribute part of my success to smoking marihuana. It makes me feel as if I am living with AIDS rather than just existing. My appetite returns, and once I have eaten, I don't feel sick anymore. Marihuana improves my state of mind, and that makes me feel better physically."
>
> (Grinspoon and Bakalar, 1993).

Whether genuinely beneficial or not, marijuana has become very popular with AIDS patients. It is estimated that 80% of the 10,000 members of the famous (or infamous) San Francisco Cultivator's Club (see Chapter 7) are AIDS patients.

Pain

As reviewed in Chapter 3, there is an increasing body of evidence from experiments in animals that activation of the cannabinoid system in the central nervous system among other things reduces the sensitivity to pain. It is possible that ongoing synthesis and release of endogenous cannabinoids in the brain play a role in setting the level of pain sensitivity at any particular time. Some have described the cannabinoid system as one that is parallel to the better known endogenous opioid system for controlling pain sensitivity (Fields and Meng, 1998). The two systems are distinct but they also overlap as cannabinoids can make morphine and other opiates more effective in relieving pain, and conversely opiates make THC more effective as an analgesic.

Clinical pain comes in many varieties from the severe but usually short-lived pain which follows injury or surgical operation to the chronic and often disabling pain which often accompanies such illnesses as rheumatism and arthritis, or cancer. As the British Medical Association Report on Therapeutic Uses of Cannabis (1997) puts it :

> Pain is perhaps the commonest of all medical symptoms requiring drug treatment.

Many different analgesic (pain-relieving) medicines are available, from aspirin and the many aspirin-like antiinflammatory drugs that act on peripheral inflamed tissues, to morphine, codeine and other opiates that act directly on the CNS. None of them are completely satisfactory. Use of aspirin-like drugs carries with it the danger of irritation and ulceration of the stomach, which can lead to dangerous internal bleeding; several thousand people die each year because of these side effects. Morphine and other opiates often cause severe constipation and at high doses they can depress respiration and cause death. The repeated use of opiates can lead to the development of tolerance, so that patients become less and less sensitive to the drugs and require increasing doses. The belief that the medical use of opiates can lead to addiction and dependence has in the past inhibited doctors from using these drugs, although it is increasingly recognized that addiction and dependence are rarely if ever encountered in the medical context. As with cannabis, the psychotropic effects of opiates are disturbing rather than pleasurable to most patients. Nowadays many patients are provided with medical devices that permit the self-administration of morphine to counteract chronic pain; they learn to titrate the dose of drug to obtain the maximum pain relief without becoming stuporous and intoxicated.

Some of the most distressing forms of clinical pain are those that arise from damage to nerves or to the spinal cord or brain. They can arise from many different causes, as a consequence of accidental or surgical injury to nerves, in patients with diabetes or AIDS, which often lead to damage in peripheral nerves, or in some forms of cancer where the tumor presses on or damages nerve fibers. A particularly disturbing syndrome is the so-called *phantom limb pain*, which occurs in as many as one-third of patients who have experienced a surgical limb amputation. They continue to experience pain arising from the damaged nerve fibers that previously had innervated the limb and the pain is experienced as though it originated from the limb that is no longer there. These so-called neuropathic pain syndromes are often long lasting and severe. They are very hard to treat, as even the most powerful analgesic drugs, the opiates, are generally ineffective. In some cases, patients respond to treatment with antiepilepsy drugs such as carbamazepine, phenytoin, or gabapentin, or to drugs used more commonly in the treatment of depression such as amitriptyline. But for many, neuropathic pain remains untreatable. Another common painful condition is migraine — a severe and disabling form of headache caused by local inflammation of the blood vessels in the membranes overlying the brain. Repeated migraine attacks occur in as many as 20% of women and 10% of men.

An encouraging feature of the results on animal models is that THC has been reported to be an effective analgesic in a model of neuropathic pain in rats, in which the sciatic nerve (which innervates the hind limb) is damaged surgically. Morphine and related opiates have previously been shown to be ineffective in this animal model. The historical literature on the medical uses of cannabis has also long stressed its value in the treatment of a variety of painful conditions. During the nineteenth century, cannabis was the drug of choice for the treatment of migraine (for review see Russo, 1998).

Given this background one might expect to find that the treatment of clinical pain was one of the most important uses of cannabis-based medicines. However, the results available so far from clinical trials are fairly disappointing. The studies are reviewed in the British Medical Association (BMA) Report, and in the NIH and American Medical Association (AMA) Reports. Double blind–placebo controlled studies with dronabinol in single doses from 5 to 20 mg in a total of 46 cancer patients showed significant pain relief 3–6 hours after drug administration, but only at the highest doses tested (15 and 20 mg). The 20 mg dose of dronabinol was equivalent to 120 mg of codeine, but it caused unpleasant and alarming psychic effects (e.g., depersonalization, loss of control) in most patients. Another double blind–placebo controlled study of the synthetic cannabinoid levonantrodol given by intramuscular injection to 56 cancer patients showed significant pain relief that persisted for up to 6 hours after the highest drug doses (2.5 and 3 mg). Drowsiness was a common side effect but few other psychoactive effects were reported. The only other positive results came from controlled clinical studies conducted with single patients treated in double blind–placebo controlled studies. In one patient suffering from a painful spinal cord injury, dronabinol (5 mg) was found to be equivalent to 50 mg codeine in alleviating pain, but with the advantage over codeine of relieving spasticity (muscle spasm). Another patient with severe chronic pain of gastrointestinal origin (familial Mediterranean fever) was treated with placebo or with cannabis oil (plant extract dissolved in olive oil). The patient's demand for morphine was significantly lower during treatment with cannabis then when receiving placebo. In contrast to these positive results, a trial of intravenously injected dronabinol in 10 healthy patients who had undergone wisdom tooth extraction yielded only equivocal results, with little indication of any significant pain relief. This may simply mean, however, that this type of acute inflammatory pain is not the one most likely to respond to cannabinoid therapy. It is unfortunate that no clinical studies yet exist examining the usefulness of cannabinoids in treating pain of neuropathic origin. Some patients with phantom limb pain have reported beneficial effects after taking cannabis, and it is possible that cannabis-based medicines might help this group whose pain is particularly hard to treat with conventional medicines. Despite the earlier popularity of cannabis in the treatment of migraine, there have been no controlled trials in this condition. There are, however, anecdotal reports:

Carol Miller, who suffers from migraine, describes her experiences as follows:

". . . it wasn't until college that I was given the diagnosis of migraine and received medication. The college infirmary prescribed [coated aspirin], which helped somewhat with the headache but not with visual effects or the nausea. It also gave me tremendous heart-burn.

One time the pain was so severe they gave me an injection of [a synthetic opioid], which pretty completely wiped out the pain but left me very light-headed.

. . . Several years later the migraine returned, and my husband said he had read that marihuana was good for headaches. I was amazed. Two hits and a short rest completely warded off the nausea and headache. As soon as I noticed flickering visuals that forewarned me of an approaching migraine, I could take a little cannabis and a short nap and the migraine would not develop at all. I was usually ready to go back to work in half an hour. It gave me a feeling of tremendous power to be finally in such control of my migraine.

In the eighteen years since I began using cannabis to relieve migraines, I have been caught away from home several times without my herb. Once I tried taking Tylenol and found it helped a little with the pain but not at all with the nausea, or the visual effects."

(Grinspoon and Bakalar, 1993)

Although there are no reports of controlled trials with nabilone in the treatment of pain, Dr. W. Notcutt, who runs a Pain Relief Service in Yarmouth, England has used the drug in some 60 of his patients (House of Lords Science & Technology Committee Report, 1998). The patients suffered from chronic untreatable pain arising from a variety of causes; these included multiple sclerosis, peripheral nerve damage, cancer, and back pain. Nabilone, given in doses up to 3 mg per day, provided useful pain relief for 18 patients, 15 had equivocal results or were unable to tolerate the drug, and 27 obtained no benefit. While a 30% success rate might not seem impressive, these patients represented the worst problems of the Pain Relief Service, and were reported not to respond to conventional analgesics or to placebo. The adverse side effects were drowsiness and unpleasant psychic effects, and these were sufficiently severe to lead several patients to abandon further use of the drug despite having obtained some pain relief. On the positive side, in addition to pain relief some patients reported that nabilone improved their sleep, relieved muscle and bladder spasms, relieved constipation, and led to increased relaxation and relief of anxiety and depression. A placebo-controlled trial of nabilone in the treatment of pain has not yet been carried out.

The usefulness of cannabinoids in the treatment of pain appears to be limited. When given by the oral route the usual problems of variable and delayed absorption are compounded in this context by the very narrow window that seems to exist between a pain-relieving and an intoxicant dose of cannabinoid. However, it is worth bearing in mind that morphine and related opiate drugs are also powerful intoxicants, yet they also play a crucial role in the treatment of clinical pain despite the narrow window that exists between pain-relieving and intoxicant doses. Although the usefulness of smoked marijuana in alleviating clinical pain is based only on anecdotal reports from individual patients, it is not difficult to believe that this route of administration has many advantages over the oral route. Smoking delivers THC to the circulation rapidly—like an intravenous injection. It thus permits a more precise titration of plasma THC levels allowing some patients to achieve the desired effect of pain relief while avoiding unpleasant psychic side effects. Smoking marijuana could be compared with a patient-controlled analgesia morphine pump with which patients self-administer morphine to the desired levels. As with many of the other potential medical indications for cannabis, there have been no controlled trials with smoked cannabis, although there are many individual anecdotes:

Lynn Hastings used cannabis to alleviate pain and spasms resulting from her rheumatoid arthritis. The following extract is from a sworn affidavit filed with the Court when she was tried for growing cannabis plants in 1989:

> *"There is not any cure for my disease. I must do what I feel is right and safe for me. I want to live long enough to enjoy my grandchildren and have the most fulfilling life that I am able to have.*
>
> *Marijuana has helped me very much. When I smoke a marijuana cigarette I receive instant relief from my pain. I am able to concentrate on my muscle spasms and relax the area that is giving me the most pain. The marijuana affects me like a wave of relief. The pain relief is fast and the effect on my mind is only for a couple of hours. I am able to think clearly without a drug hangover the next day.*
>
> *The fatigue that I feel from my juvenile rheumatoid arthritis is also remedied within half an hour. I am able to do my housework or cook my family's dinner. I am able to talk to people and have a normal conversation with friends, family or phone calls. Another benefit from the marijuana is that it helps me with my depression.*
>
> *. . . I never felt that I needed the marijuana other than for pain and muscle spasm, nor did I need to increase the amount of marijuana to obtain the pain relief that I required."*
>
> (Randall, 1991)

Multiple Sclerosis

Multiple sclerosis (MS) is the most common disabling nervous system disease of young adults, with an estimated 85,000 living with the condition in the United Kingdom and more than 250,000 in the United States. Multiple sclerosis is a progressive, degenerative disease, in which the brain and the spinal cord nerves are damaged by the gradual destruction of myelin, the protective, insulating layer of fatty tissue that normally coats nerve fibers. The precise cause of the disease is not known, but it is thought to represent an autoimmune condition, in which the body's immune system becomes inappropriately sensitized to some component of myelin — leading to its attack and progressive damage by the immune system. The disease usually progresses in stages, with periods of remission between, but it is ultimately life threatening. The symptoms are variable, depending on which particular nerves or regions of CNS are damaged, but it often manifests itself with symptoms of muscle spasticity, pain, and bladder and bowel dysfunction. The British Multiple Sclerosis Society reported the results of a survey of their 35,000 members to the House of Lords Science & Technology Committee Cannabis Report 1998. Fatigue was the most frequent symptom reported by 95% of patients, followed by balance problems (84%), muscle weakness (81%), incontinence (76%), muscle spasms (66%), pain (61%), and tremor (35%).

There are several medicines available to treat the symptoms of MS, but none are wholly effective. The drugs baclofen and diazepam (Valium®) help to relax muscle spasms by activating receptors for the inhibitory chemical messenger molecular GABA in brain and spinal cord, thus counteracting overactivity in the flow of excitatory signals to muscles. Both drugs may cause side effects, including sedation, drowsiness, and confusion. Dantrolene acts directly on the muscle to dampen the force of contraction but can cause serious side effects (headache, drowsiness, dizziness, malaise, and nausea). Oxybutynin, flavoxate, and propantheline can be helpful in controlling urinary incontinence; they block the actions of the chemical signal acetylcholine that trigger bladder emptying. All of these drugs may cause dry mouth, blurred vision, constipation, and difficulty in initiating urine flow. Chronic pain in MS sufferers is often hard to treat, but drugs used in the treatment of epilepsy (carbamazepine, phenytoin) or depression (amitriptyline) and even opiates are sometimes used. Despite the relatively large numbers of patients with MS it has not attracted much attention from pharmaceutical research companies. The muscle relaxant tizanidine launched in the United States in late 1997 was the first new drug to receive approval for the treatment of muscle spasticity in 20 years.

Multiple sclerosis represents a promising target for cannabis-based medicines (Clifford,1983; Consroe and Snider, 1986; Randall, 1991). Illegal self-medication with cannabis both in smoked and oral forms is already common. The United Kingdom MS Society estimates that as many as 4% of their members are treating their symptoms with cannabis. Anecdotal reports suggest that cannabis can relieve not only the muscle spasms and the pain, but in some patients it can also improve bladder control. The sedative properties of cannabis may also offer sound sleep to patients whose sleep is otherwise frequently disturbed by painful muscle spasms and the frequent need to urinate. There is a substantial historical tradition for the use of cannabis in the treatment of various types of painful muscle spasms. This also has a sound scientific rationale. Cannabinoid CB1 receptors are found in particularly high density in those regions of the brain that are involved in the control of muscle function — the basal ganglia and the cerebellum. The receptors are densely located on output neurones in the outflow relay stations of the basal ganglia (substantia nigra pars reticulata and globus pallidus) where they are well placed to affect the control of movements. Activation of the cannabinoid receptors is known to suppress movements and can lead to a condition of catalepsy, in which the person or animal may remain conscious but immobile for considerable periods. It is not surprising, therefore, that cannabinoid drugs possess antispastic properties. In an animal model of MS in mice (allergic encephalomyelitis) the animal's immune system is sensitized to a component of its own myelin and there is progressive nervous system damage accompanied by muscle tremor. This and other symptoms in this animal model can be suppressed by treatment with THC. In this model, repeated treatment with THC also has the effect of slowing down the development of the syndrome — suggesting that cannabinoids might even be able to alter the course of an autoimmune disease, perhaps because of their ability to dampen immune system activity.

Multiple sclerosis sufferers are clearly in need of better treatment, and the rationale for using cannabis appears sound, so it is disappointing to find so few properly controlled clinical trials of cannabis or cannabinoids for this condition. There have been only six published clinical trials, involving a total of only 41 subjects worldwide, and the results have been equivocal (for review see British Medical Association Report, 1997). In one double blind–placebo controlled study, nine patients received single doses of 5 or 10 mg of dronabinol orally. Rated objectively, both doses caused a significant improvement in spasticity. The improvement peaked at around 3 hours post drug and lasted for a few hours. Only three of the nine patients, however, felt subjectively that the drug had improved their condition — they felt "loose and better able to walk." Another study in 12 MS patients given 7.5 mg of dronabinol found no changes in objective ratings of weakness, spasticity, coordination, gait or reflexes, but in this study most patients noted significant subjective improvement. Adverse effects were common and most of the patients did not request further treatment with dronabinol. In a study of smoked cannabis involving 10 normal volunteers and 10 MS patients some patients noted subjective improvements. Objective measurements of posture and balance, however, showed that the drug impaired these functions in both normal subjects and in the MS patients, with the latter showing the greatest degree of drug-induced impairment. An open trial in eight MS patients showed mild subjective (but not objective) improvements in tremor and in feelings of well-being in five patients who received 5–15 mg dronabinol every 6 hours and subjective and objectively measured improvement in two patients. One of these showed a remarkable improvement in tremor and hand coordination (Fig. 4.2). All the patients reported feeling high and two found this unpleasant. There have also been reports on single patients. One MS patient showed dramatic improvement in tremor and coordination after smoking a single marijuana cigarette. In another study an MS patient took nabilone or placebo on a double-blind basis. Nabilone (1 mg per day) produced a clear improvement in muscle spasms, control of night-time urinary function, and feelings of well-being (Martyn et al., 1995; Fig. 4.3).

The remainder of the evidence supporting the use of cannabis in MS is anecdotal. In an attempt to make this more systematic, Consroe and colleagues (1996) undertook a survey of MS patients in the United Kingdom and the United States who admitted to self-medicating with smoked marijuana. Their report summarizes the responses received to an anonymous postal questionnaire from 112 MS patients. The average age of the subjects was 44 years, and they had suffered from MS for an average of 14 years. They had smoked marijuana for an average of 6 years (on average 3 times a day for 5–6 days each week). Although the

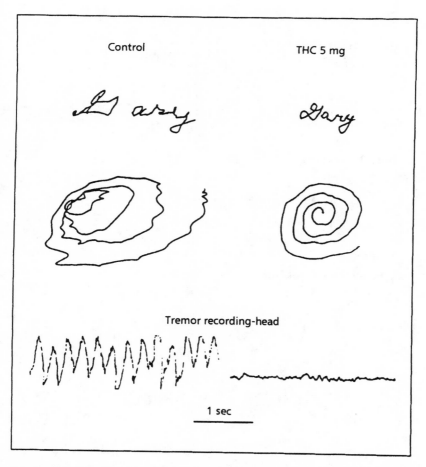

Figure 4.2. Handwriting and head tremor recorded before and 90 minutes after ingestion of 5 mg of THC in a patient with multiple sclerosis. From Clifford (1983). Reprinted with permission from Lippincott Williams & Wilkins.

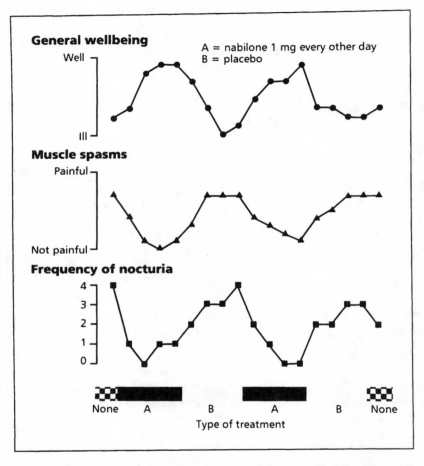

Figure 4.3. Double-blind, placebo controlled crossover study of nabilone in a patient with multiple sclerosis. The patient took 1 mg nabilone every other day for two periods of 4 weeks, alternating with 4-week periods of placebo. Neither patient nor doctor knew when the active drug was being given. There was a clear improvement in symptoms during the periods of treatment with nabilone. Data from Martyn et al. (1995) redrawn by British Medical Association (1997) and reprinted with permission from Gordon and Breach Publishers.

great majority of patients responding to the questionnaire reported that they benefited from smoking cannabis, the results of such surveys need to be treated with considerable caution. As the authors themselves caution, previous research has shown that more than 65%–70% of patients with MS report improvements when given a completely ineffective placebo. There is an unmet medical need and patients want new medicines to work. The results of this survey were, nevertheless, impressive. If one takes only those symptoms in which improvement was reported in more than 70% of those surveyed, there are some startling figures. More than 90% of the respondents reported that marijuana improved their spasticity at sleep onset (96.5%), their muscle pain (95%), their night-time spasticity (93%), their leg pain at night (92%), their depression (90%), and their tremor (90%). Between 70% and 90% reported improvements in anxiety, daytime spasticity, tingling, numbness, facial pain, muscle weakness, and weight loss. Smaller proportions reported some benefit to bladder or bowel function or improvements in impaired vision, balance, speech, or memory.

The reports from individual patients are often moving. Clare Hodges (not her real name) is an MS sufferer who has been self-medicating with cannabis for more than 6 years. She has formed an organization called Alliance for Cannabis Therapeutics (ACT) in the UK to help fellow sufferers and to campaign for cannabis to be made available for medical uses. She gave the following evidence to the House of Lords Science & Technology committee inquiry into cannabis (1998):

I am a 40-year-old married woman with two children and I first developed multiple sclerosis when I was 25 years old. I started by experiencing spasticity, nausea and loss of sensation. Over the years I have developed further problems of more sickness and stiffness, poor appetite, and great discomfort in my bladder. This discomfort resulted in me going to the toilet at least 12 times a night and not sleeping. This in turn exacerbated other symptoms of impaired vision, poor balance, fatigue and susceptibility to infection.

I had been prescribed many different medicines for the various problems. None of these medicines provided sustained relief. Baclofen, Valium and ACTH all had unacceptable side effects in my case. Temazepam did help me sleep but made me anxious and I found it very habit-forming. In particular I sought relief for the bladder problem. I was not helped by imipramine, desmopressin or oxybutynin, and took a daily dose of nitro-furantoin to prevent constant urinary tract infections.

I had been treated for nine years with orthodox medicines, but as I had not been able to find relief for my bladder, I decided to treat myself. I read an article in an American journal about how some doctors report that people with MS benefit from using cannabis. . . . A friend got hold of some cannabis for me and showed me how to smoke it with tobacco. It was effective in about five minutes. The tension in my bladder and spine melted away, and I felt less sick and stiff. I could move and generally function with greater ease and I slept soundly that night without medication.

After few months of taking cannabis, I found I was able to reduce the doses of my prescribed medication, so that I did not take anything for my bladder or to help me sleep. I no longer take daily antibiotics, as I do not have such problems emptying my bladder. Cannabis has not "cured" my MS, but my general health is very much improved. As well as relieving specific acute problems such as bladder dysfunction, discomfort in my spine, and nausea, it has helped long term problems such as poor sleeping and lack of appetite. My MS symptoms vary considerably. Sometimes I appear quite able bodied for short periods, and at other times I look and sound very handicapped. Similarly I can be cheerful about my situation, but when the MS is bad, I become very introspective and negative. I known a lot of the therapeutic benefit is psychological as well as physical. MS makes me slightly under par all the time, so that even the simplest task takes an enormous effort and leaves me exhausted. I don't have to get "high" for cannabis to lift my mood, make me feel calm and positive and able to carry on more normally. This kind of mood-altering effect often seems to be desirable in serious, chronic illnesses judging by the large numbers of people with MS who are prescribed anti-depressants and tran-quillisers.

The relief cannabis gave me has been sustained over the years, the main improvement being bladder control. It has not helped all the symp-toms I have. For example, it has not affected my impaired vision. I do not feel addicted to cannabis any more than any ill person is "addicted" to their medication. If, for some reason (e.g. cost or availability) I cannot take cannabis, I do not "crave" it in any way nor experience withdrawal. All that happens is that my MS symptoms return."

Elsewhere in her evidence to the Committee she said:

> I know that in large amounts cannabis can make you very uncoordinated and clumsy rather like multiple sclerosis can and in large amounts it can make you heady and drunk. I have experienced that a couple of times and I do not like it, so I do not do that. That it is not why I take it. The reason I took it was to help my physical symptoms.

The Institute of Medicine report (1999) included the following quote from B. D., who spoke at one of the Institute workshops held in Louisiana. She is one of the eight patients who are legally allowed to smoke marijuana under a Compassionate Use Protocol in the United States. She uses marijuana to relieve nausea, muscle spasticity, and pain associated with multiple sclerosis:

> . . . When I found out that there was a program to get marijuana from the government I decided that was the answer. I was not a marijuana smoker before that. In fact, I used to consider the people I knew who smoked marijuana as undesirables. Now, I myself am an undesirable.
>
> But it works. It takes away the backache. With multiple sclerosis, you can get spasms, and your leg will just go straight out and you cannot stop that leg. You may have danced all your life and put the leg where you wanted it to be, but the MS takes that from you. So I use the swimming pool and that helps a lot. The kicks are much less when I have smoked a marijuana cigarette. Since 1991, I've smoked 10 cigarettes a day. I do not take any other drugs. Marijuana seems to have been my helper. At one time, I did not think much of people who smoke it. But when it comes to your health, it makes a big difference.

In addition to MS, marijuana is also used illegally by other groups of patients who suffer from disabling illnesses that are accompanied by painful muscle spasms. These include cerebral palsy, torticollis, various dystonias, and spinal injury. P. Consroe, who conducted a postal survey of the use of marijuana by MS patients, has carried out a similar survey of patients suffering from spinal injuries. He reported the results obtained from 106 such patients at the International Cannabinoid Research Society Symposium in 1998. The patients smoked an average of four joints a day, 6 days a week and had been doing so on average for more than 10 years. More than 90% reported that marijuana helped improve symptoms of muscle spasms of arms or legs, and improved urinary control and function. As with the previous survey on MS patients such results need to be viewed with caution, nevertheless, they seem quite impressive. At least such surveys may help to pinpoint the relevant symptoms to focus on as outcome measures in future controlled clinical trials of cannabis or cannabinoids.

Glaucoma

Glaucoma is a disease in which the fluid pressure within the eye (intraocular pressure [IOP]) becomes abnormally high, possibly because of obstruction to the outflow of fluid from the eye. Over time, this damages the optic nerve, causing progressive loss of sight and eventually blindness. Some patients suffer painful acute attacks with severe headaches and vomiting; in others, visual problems such as halos and blind spots are the most prominent symptoms. Glaucoma is the leading cause of blindness in the Western world, affecting an estimated 2 million Americans over the age of 35 and similar numbers elsewhere.

A variety of medicines are available to treat glaucoma, these act principally either to reduce the rate of fluid formation in the eye or to increase the drainage of fluid away from the eye. Usually the medicine is administered directly to the eye by means of eye drops or ointment, as this permits a much smaller dose of the compound to be administered, thus avoiding many of the adverse side effects seen if the same medicines are given by mouth. Effective medicines include the acetylcholine-like drugs pilocarpine or carbachol, which stimulate fluid outflow, the noradrenaline beta-receptor blocker timolol, and alpha$_2$-adrenoceptor drugs that reduce fluid formation in the eye, prostaglandin analogues that increase fluid outflow, and inhibitors of the enzyme carbonic anhydrase that is involved in ocular fluid formation. Topically applied carbonic anhydrase inhibitors are a recent advance, as previously such compounds were given by mouth and tended to cause a number of adverse side effects. Most of these drugs are efficacious, have few adverse side effects, and possess long duration of action, but none of them are ideal. Although it is possible to use these drugs to lower IOP and to delay the course of the disease, it is hard to maintain a lowered IOP for 24 hours 7 days a week—which is ideally what is required. On repeated use of any of the available medicines patients tend to become tolerant to them and they become less and less effective.

It was therefore of considerable interest to find that cannabis had the ability to lower IOP. This was discovered almost by accident by Dr. Robert Hepler and his colleagues at the University of California Los Angeles in the early 1970s. They were looking for effects of marijuana on the eye that might help the police to identify illicit drug users, and found that smoked marijuana caused a significant reduction in IOP of around 25%. This finding has since been confirmed in several other studies using both smoked and orally administered cannabis. Synthetic cannabinoids given intravenously were also shown to lower IOP, and only the psychoactive compounds that have high affinity for CB1 receptor (Δ^9-THC, Δ^8-THC, and 11-hydroxy-THC) were active, while cannabidiol and cannabinol were only weakly active. The results in human volunteers were also repeated successfully in numerous animal experiments (for review see Adler and Geller, 1986).

Although these findings were followed by a number of anecdotal reports of the beneficial effects of marijuana in glaucoma patients there have been disappointedly few controlled clinical studies in such patients. In the best double-blind placebo cigarette-controlled trial, with a crossover design, Merritt et al. (1980) studied 18 patients who smoked cigarettes containing 2% THC. There was a significant reduction in IOP (Fig. 4.4) but this was also accompanied by a number of undesirable side effects, including lowered blood pressure and psychic effects. In addition, cannabis and cannabinoids cause other ocular effects, including reddening of the eye and surrounding conjunctival tissues because of dilated blood vessels, and decreased tear formation leading to a dryness of the cornea that can be damaging. Keith Green, a professor of ophthalmology at the Medical College of Georgia in the United States told the House of Lords Science and Technology Committee Cannabis inquiry (1998) of the results of his own studies in more than 300 human subjects with both normal and raised IOP. Cannabis caused an average of 25% decrease in

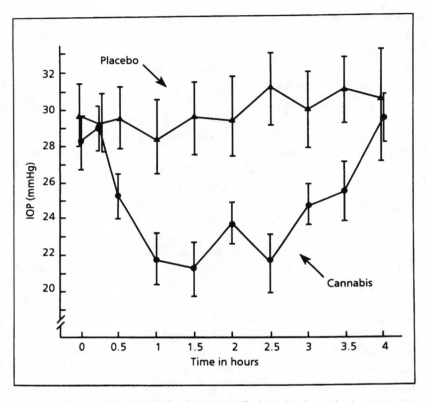

Figure 4.4. Double-blind placebo controlled trial of smoked marijuana (2% THC content) versus placebo cigarettes in 18 patients with glaucoma, measuring intraocular pressure (IOP). Results are mean values ± standard error. From Merritt et al. (1980). Reprinted courtesy of *Ophthalmalogy*.

IOP, which lasted for 3–4 hours. He concluded, as have most other ophthalmologists, that treatment of glaucoma with cannabis is impractical. Patients would have to smoke marijuana several times a day in order to maintain lowered IOP, a regime that would not be practicable in view of the psychoactive effects of the drug and its effects on cognitive function. There are no reports of controlled studies involving the long-term use of marijuana in glaucoma patients, so it is impossible to assess whether the drug does alter the course of the disease. The Institute of Medicine Report (1999) came to the following negative conclusion:

> Although glaucoma is one of the most frequently cited medical indications for marijuana, the data do not support this indication. High intraocular pressure (IOP) is a known risk factor for glaucoma and can, indeed, be reduced by cannabinoids and marijuana. However, the effect is too short-lived, requires too high doses, and there are too many side effects to recommend lifelong use in the treatment of glaucoma. The potential harmful effects of chronic marijuana smoking outweigh its modest benefits in the treatment of glaucoma.

Nevertheless, the effects of cannabis on IOP are intriguing and worthy of further study. Some of the clinical studies with cannabis in glaucoma patients suggest that its ability to lower IOP is additive to the effect of some of the other conventional treatments — indicating that it may act through a completely different mechanism. Until recently it was thought possible that the effect of cannabis in IOP might be explained by the effect of the drug in lowering blood pressure, which in turn would lead to decreased fluid formation in the eye. It is now clear, however, that cannabinoid CB1 receptors do exist in the eye, making it more likely that a direct effect of cannabis locally in the eye is responsible for the lowering of IOP. Initial attempts to administer THC topically to the eye by means of eye drops yielded equivocal results and the poor water solubility of THC makes it difficult to administer in this way. D. Pate et al. (1998) obtained promising results with a metabolically stable analogue of anandamide and the synthetic cannabinoid CP-55,940 given topically to rabbit eyes, using a novel solubilizing agent called beta-cyclodextrin. Both caused a lowering of IOP that could be blocked by the CB1 antagonist SR141716A. If cannabis-based medicines could be given successfully by topical application to the eye, this would avoid many of the undesirable side effects associated with smoking or oral administration.

In order to assess the potential of cannabis in the treatment of glaucoma much more needs to be known about the mechanism by which it acts on IOP. It is also not clear whether the effects of cannabis are additive to the newer highly effective treatments that are available, the top-

ically applied beta-blockers, alpha$_2$-adrenoceptor drugs, carbonic anhydrase inhibitors, and prostaglandin analogues. As the NIH *Report on
the Medical Uses of Marijuana* (1997) concluded:

> If marijuana were not to be additive to one of these agents, marijuana
> would be obsolete, since these agents have no systemic side effects (other
> than slightly dry mouth in some patients with apraclonidine and bro-
> monidine), and they have a duration of action of 12 to 24 hours.

Although glaucoma does not appear to be a particularly good target
for cannabis-based medicines, some individual patients have reported
that smoking marijuana dramatically improved their symptoms. Perhaps
the most famous case was the American, Robert Randall who began
treating himself with smoked marijuana in the 1970s, after he had lost a
substantial degree of vision. He gives the following account:

> Despite my use of every pharmaceutical agent in the inventory, my eve-
> nings were routinely visited by tricolored halos — a signature of ocular pres-
> sure over 35 mm HG (millimetres of mercury). On some nights the halos
> were muted. On other evenings they appeared as hard crystal rings emanat-
> ing from every source of light. And then there were nights, not so rare, of
> white-blindness — the world rendered invisible by its brilliance. Clinical
> translation: ocular tension in excess of 40 mm Hg. To summarize, things
> were not going very well.
>   Then someone gave me a couple of joints. Sweet weed! That night I
> made and ate dinner, watched television. My tricolored halos arrived,
> which made watching TV less interesting. So I put on some good music,
> dimmed offending lights, and got into some serious toking. I happened to
> look out of my window at a distant street light and noticed what was not
> there. No halos. That's when I had the full blown, omni-dimensional tech-
> nicolor cartoon light-bulb experience. In a transcendant instant the spheres
> spoke! So simple. Old messages — new context. You smoke pot, your eye
> strain goes away.
>   . . . My ever-eroding visual fields stabilized. My slide into darkness
> slowed, then halted. As my glaucoma came under medical management,
> other aspects of life began to right themselves. I escaped welfare and took a
> part time teaching job at a local college.

(Grinspoon and Bakalar, 1993)

Randall was determined to have access to marijuana to treat his glaucoma. After losing his home-grown cannabis plants in a police raid, he launched a lawsuit against the United States government, and in 1978 won a settlement in which the government agreed to supply him with National Institute on Drug Abuse (NIDA)-grown marijuana cigarettes free of charge under a compassionate FDA investigational drug protocol, at a rate of about 300 cigarettes a month. In effect Randall was considered to be a research subject in an FDA-approved single-person drug trial, a process involving much paperwork and red tape for Randall's doctor. This set a precedent, and eventually this concession was extended to 30 patients in the United States, until the Bush government terminated the program in the 1980s. Even now, the United States government continues to supply free marijuana to some eight remaining patients, including Randall.

Epilepsy

Epilepsy is a common nervous system disorder that affects about 1% of the population. Patients experience repeated periods of uncontrolled electrical activity in the brain. In the more severe forms of epilepsy this can lead to a *fit* in which the patient becomes unconscious and experiences muscular convulsions, or in *petit mal epilepsy*, the seizure may be milder and lead only to a temporary lapse in consciousness, often lasting only a few seconds. There are several drugs available that help to control epilepsy and lessen the risk of seizures, these include carbamazepine, sodium valproate, phenytoin, phenobarbitone, primidone, ethosuximide, and clonazepam. In the past few years a new generation of antiepileptic drugs has also been introduced: vigabatrin, lamotrigine, gabapentin, and topiramate.

Although none of these medicines is devoid of adverse side effects, it is usually possible to find the medicine that is best tolerated for an individual patient and to effectively control their symptoms. There is a substantial number of epileptic patients, however, whose epilepsy is not well controlled by drugs and many of these are severely and permanently disabled — being unable to go to work or to drive a car.

Cannabis was commonly used in the nineteenth century to treat epilepsy, but there has been little interest more recently since the anti-epileptic drugs described above became available in the 1930s and subsequently. Animal data do show antiseizure activity of THC in some experiments, but there are also conditions in which cannabinoids can make animals more susceptible to seizure activity. An interesting observation is that the nonpsychoactive cannabinoid cannabidiol also appeared to be active as an antiseizure compound in some animal studies. There have been very few clinical studies with this compound, but one placebo-controlled trial in 15 treatment-resistant epileptic patients suggested that cannabidiol in doses of 200 or 300 mg by mouth might have beneficial effects. Attempts to follow up these potentially useful findings, however, failed to confirm any positive effect of this dose of cannabidiol on seizure frequency, although a single patient treated with a higher dose of cannabidiol (900–1200 mg per day) seemed to benefit (Hollister, 1986; British Medical Association Report, 1997). Cannabidiol has no appreciable activity at either of the known cannabinoid receptors, so if it is active as an anticonvulsant this must presumably involve an action at some hitherto undiscovered cannabinoid receptor.

Anecdotal reports of the use of smoked marijuana by epileptic patients are mixed. Some patients claim to derive significant benefits and were able to reduce the doses of conventional antiepileptic drugs needed. At least one case has been reported, however, in which oral THC (20 mg) precipitated a seizure in a patient with a previous history of epilepsy. Other reports suggest that cannabis may counteract the effectiveness of drugs used in the treatment of petit mal epilepsy. Even the most ardent advocates of the medical uses of marijuana caution:

> Cannabis is by no means a cure-all for epilepsy. . . . Epileptics who are interested in trying cannabinoids should be careful about oral THC. Those who use cannabinoids should be aware that they may become more susceptible to seizures when they withdraw from treatment.
>
> (Rosenthal et al., 1997)

It is difficult to see epilepsy as a high priority area for research on the potential use of cannabis or cannabinoids. The introduction of sev-

eral new antiepileptic medicines in the past few years means that clinicians in this field are still finding out how best to use these new drugs,
and discovering to what extent they may be effective in patients who
were previously resistant to antiepileptic medication. There will still undoubtedly be some patients whose symptoms will not be well controlled
by any of the existing medicines. It is possible that cannabis might benefit these patients. More research on the nonpsychoactive cannabidiol
would also appear to be well justified.

Bronchial Asthma

Asthma, an illness that involves a chronic inflammatory state of the airways in the lungs, is a common disease that is increasing rapidly, especially among the young, in the Western world. Sufferers experience periods of wheezing and difficulty in breathing that can become disabling
and in extreme cases life threatening. Fortunately a number of effective
medicines are available, most of which are administered directly to the
lungs by inhaling a standard dose by means of a metered aerosol device.
Beta-adrenoceptor agonists, compounds that mimic the actions of the
naturally occurring chemical messenger noradrenaline, act directly to relax the bronchial tubes and rapidly make breathing easier. The first of
these to be used was isoprenaline, but this can have harmful stimulant
effects on the heart and has been replaced by safer and more selective
beta agonists such as salbutamol and long-lasting drugs such as sal-
meterol that only need to be taken once or twice a day. The beta agonists
treat the acute symptoms of the illness, but inhaled steroids, such as
beclomethasone and fluticasone have had a major impact on the treatment of asthma as these drugs treat the inflammatory condition itself and
help to keep it under control. Another addition to the antiasthma medicine cabinet recently has been the drug montelukast, a compound taken
by mouth that has powerful antiinflammatory actions in the lung.

The possible use of cannabis in the treatment of asthma arose from
studies of the effects of marijuana on respiratory function in normal
healthy volunteers and in asthmatic subjects undertaken in the 1970s
(Hollister, 1986). A fall of almost 40% in airway resistance was observed

in volunteer studies. This led to a number of studies of smoked marijuana and oral THC in asthmatic subjects during the 1970s — a period before the modern antiasthma medicines had become available. In acute studies in 14 asthmatic subjects smoked cannabis was found to cause a bronchodilation comparable to the then standard inhalation drug isoprenaline. However, smoked marijuana is clearly not suitable for long-term use in asthmatic subjects because of the irritant effects of various components present in the smoke. Oral THC was found to be impractical, as the doses needed for bronchodilation were clearly psychoactive. The most interesting approach has been to devise methods for administering THC directly to the lungs by means of an aerosol. In one placebo-controlled study in 10 asthmatic subjects a THC aerosol that delivered 200 micrograms of THC was compared with a salbutamol aerosol (100 micrograms). Both drugs significantly improved respiratory function, the effect of THC was slower in onset but reached a similar maximum after 1 hour (Williams et al., 1976; Fig. 4.5). However, in other studies with inhaled THC some patients found the aerosol caused irritation to the lung, chest discomfort, and coughing. This inhibited further development of this line of research, although it is not clear whether the irritant effects of the aerosol THC were due to the THC itself or to the solvents used to dissolve it.

Unless the problems of delivering THC or other cannabinoid directly to the lung can be solved there seems little future in the use of cannabis-based medicines in the treatment of bronchial asthma. There are now several powerful antiasthma medicines available. Cannabis, like the betaagonists, treats the symptoms of bronchoconstriction but there is no evidence that it alters the underlying condition. What is needed in the future are more medicines that act on the disease process itself, perhaps helping to prevent the disease's developing into its more severe forms.

Mood Disorders and Sleep

Cannabis has been advocated as a treatment for depression, anxiety, and sleep disorders. One of the first recommended uses of cannabis in West-

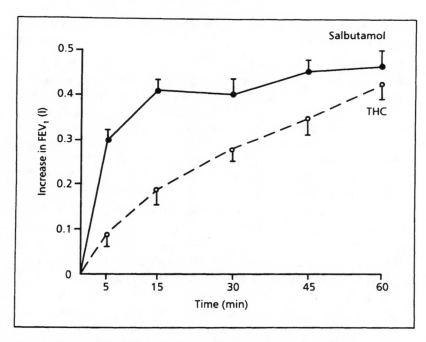

Figure 4.5. Double-blind placebo controlled trial of the bronchodilator effect (increased FEV_1) of 100 μg salbutamol (filled circles) and 200 μg THC (open circles) inhaled as a metered dose aerosol in ten asthmatic patients. From Williams et al. (1976). Reprinted with permission from BMJ Publishing Group.

ern medicine was for the treatment of depression and melancholia, and before the discovery of modern antidepressant drugs cannabis continued to be used in this way during the first half of the twentieth century. However, the few clinical trials that have been conducted with THC or nabilone in the treatment of depression or anxiety have had mixed results. Although some patients reported improvements, others found the psychic effects of the cannabinoids unpleasant and frightening. Rather than relieving anxiety, the acute effect can be to provoke anxiety and panic in some subjects — particularly those who have had no previous exposure to cannabis. Nevertheless some depressed patients report great benefits from cannabis:

> In the spring of 1990 I smoked marihuana for the first time since 1973. To my amazement, a quarter of a joint changed my self perception to match the person others saw. It was like night and day. I had experienced a similar change only a few times before, when [amitriptyline] kicked in and lifted me out of the depths. But with [amitriptyline] it took four days of rapidly increasing doses; with marihuana it took less than five minutes, every time. Since then I have been using marihuana to think clearly, to concentrate, and simply to enjoy the beauty of the world in a way I couldn't for years.
>
> (Grinspoon and Bakalar, 1993)

In sleep laboratory studies orally administered THC at doses of 10–30 mg has been shown to cause increases in deep slow wave sleep, but at the same time—as with other hypnotic drugs—there is a decrease in dreaming or rapid eye movement (REM) sleep. After repeated treatment with large doses of THC there was evidence of some degree of hangover during the morning after treatment, and a rebound in the amount of REM sleep. THC thus does not appear to offer any advantages over existing sleeping pills, and it has the disadvantage of causing intoxication prior to sleep.

This seems to be another example in which the original case for using cannabis has been made obsolete by the development of modern antidepressants (e.g., Prozac®), antianxiety agents (e.g., Valium®), and sleeping pills (e.g., temazepam and zopiclone).

Conclusions

There are clearly several possible therapeutic indications for cannabis-based medicines, but for most of them evidence for the clinical effectiveness of the drug is woefully inadequate by modern standards. The example of the use of cannabinoids in the treatment of sickness associated with cancer chemotherapy shows, however, that it is possible to amass such evidence provided resources are devoted to such studies, and unambiguous clinical trial designs are used. One of the obvious complications in the medical use of cannabis is that the window between its therapeutic effects and the cannabis-induced high is often narrow. As the Institute of Medicine report (1999) points out, however, this can sometimes be beneficial to the patient. Older patients with no previous experience of cannabis may find the psychological effects of the drug disturbing and unpleasant. But in some conditions the antianxiety effects of cannabis can have a beneficial effect, since anxiety itself tends to make the symptoms worse, e.g., in movement disorders, in cancer chemotherapy, and in AIDS wasting syndrome.

The other requirement for a human medicine is that it should be safe to use and the next chapter is devoted to that subject.

5

Is Cannabis Safe?

The initial enthusiasm for cannabis in the 1960s and early 1970s was rapidly followed by a wave of reaction in the Western world. Although scientists are supposed to try to minimize bias this has been difficult to avoid in a field so colored by issues of morality and public policy. Scientists are human beings; they may consciously or unconsciously design and manipulate research for fame and fortune (grants), and some have been guided by a moral commitment to proving that cannabis is harmful. Extravagant warnings were given, suggesting that cannabis was a highly dangerous drug that could cause chromosomal damage, impotence, sterility, respiratory damage, depressed immune system response, personality changes, and permanent brain damage. Most of these claims were later proved to be spurious and the balanced reviews by Hollister (1986, 1998) and by L. Zimmer & J. P. Morgan (1997) in their entertaining book *Marijuana Myths, Marijuana Facts* show how effectively many of them have been demolished. It is thus not necessary to deal with all of these arguments in detail here, but simply to highlight some of the factors that may determine whether cannabis is considered sufficiently safe to be reintroduced into Western medicine and ultimately whether its overall prohibition remains justified.

Toxicity

Tetrahydrocannabinol is a very safe drug. Laboratory animals (rats, mice, dogs, monkeys) can tolerate doses of up to 1000 mg/kg (milligrams per kilogram). This would be equivalent to a 70 kg person swallowing 70 grams of the drug — about 5,000 times more than is required to produce a high. Despite the widespread illicit use of cannabis there are very few if any instances of people dying from an overdose. In Britain, official government statistics listed five deaths from cannabis in the period 1993–1995 but on closer examination these proved to have been deaths due to inhalation of vomit that could not be directly attributed to cannabis (House of Lords Report, 1998). By comparison with other commonly used recreational drugs these statistics are impressive. In Britain there are more than 100,000 alcohol-related deaths and at least as many tobacco-related deaths each year. Even such apparently innocuous medicines as aspirin and related nonsteroidal antiinflammatory compounds are not safe. Thousands of people die every year because of the tendency of these drugs to cause catastrophic gastric bleeding. Hundreds more die while taking the painkiller paracetemol, because of its tendency to cause liver damage.

Long term toxicology studies with THC were carried out by the National Institute of Mental Health in the late 1960s (Braude, 1972). These included a 90-day study with a 30-day recovery period in both rats and monkeys. These studies were similar in design to those required for any new medicine before it can be approved from human use. Large numbers of animals were exposed to high doses of the drug every day, and blood samples were taken regularly to look for biochemical abnormalities during the study. At the end of the study a careful autopsy was performed on each animal, recording the weight and appearance of internal organs. Sections of the major organs were subsequently examined under the microscope to look for any pathological changes. Interestingly, these studies included not only Δ^9-THC but also Δ^8-THC and a crude extract of marijuana. Treatment of animals with doses of cannabis or cannabinoids in the range 50–500 mg/kg lead to decreased food intake and lower body weight. All three test substances initially depressed behavior but later animals became more active, and were irritable or aggressive. At the end of the study decreased organ weights were seen in the ovary, uterus, prostate, and spleen and increases were seen in the adrenals. The behavioral and organ changes were similar in monkeys but less severe than those seen in rats. Further studies were carried out to assess the potential damage that might be done to the developing fetus by exposure to cannabis or cannabinoids during pregnancy. Treatment of pregnant rabbits with THC at doses up to 5 mg/kg had no effect on birth weight and did not cause any abnormalities in the offspring. Dr. Braude concluded:

> In summary, I would like to say that Delta-9-THC given orally seems to be a rather safe compound in animals as well as in man and appears to have little teratological potential even at dose levels considerably higher than the typical human dose.

Chan et al. (1996) reported the findings of similarly detailed toxicology studies carried out with THC by the National Institute of Environmental Health Sciences in the United States, in response to a request from the National Cancer Institute. Groups of rats and mice were treated repeatedly with a range of doses of THC dissolved in corn oil, including doses many times higher than those likely to be used clinically. Each dose of the drug was administered to a separate group of 10 male and 10 female animals. In both species the doses ranged from 0 to 500 mg/kg. The animals were treated five times a week for 13 weeks, and some groups of animals were followed for a further period of 9 weeks. By the end of the study more than half of the rats treated with the highest dose (500 mg/kg) had died, but all of the remaining animals appeared healthy, although in both species the higher doses caused lethargy and increased aggressiveness. The THC-treated animals ate less food and their body weights were consequently significantly lower than those of untreated controls at the end of the treatment period, but rose back to normal levels during the subsequent recovery period. During this period animals were sensitive to touch and some exhibited convulsions. There was a tendency for the drug to cause decreases in the weight of the uterus and testes.

In further studies groups of rats were treated with doses of THC up to 50 mg/kg and mice with up to 500 mg/kg, five times a week for 2 years, a standard test to determine whether new medical compounds are liable to cause cancers. At the end of the 2 years, more treated animals had survived than controls—probably because the treated animals ate less and had lower body weights. The treated animals also showed a significantly lower incidence of the various cancers normally seen in aged rodents, in testes, pancreas, pituitary gland, mammary gland, liver, and uterus. Although there was an increased incidence of precancerous changes in the thyroid gland in both species and in the mouse ovary after one dose (125 mg/kg), these changes were not dose related. The conclusion was that there was "no evidence of carcinogenic activity of THC at doses up to 50 mg/kg." This was also supported by the failure to detect any genetic toxicity in other tests designed to identify drugs capable of causing chromosomal damage. For example, THC was negative in the so-called Ames test in which bacteria are exposed to very high concentrations of the test drug to see whether it induces any mutations. In another test, hamster ovary cells were exposed to high concentrations of the drug in tissue culture, and no effects were observed on cell division that might indicate chromosomal damage.

By any standards, THC must be considered a very safe drug both acutely and on long-term exposure. This probably reflects the fact that cannabinoid receptors are virtually absent from those regions at the base of the brain that are responsible for such vital functions as breathing and blood pressure control. The available animal data are more than adequate to justify its approval as a human medicine, and indeed it has been approved by the FDA for certain limited therapeutic indications.

Acute Effects of Cannabis

Of all the immediate actions of cannabis (Chapters 2 and 3) its psychoactive effects are undoubtedly those that give the greatest concern in considering the medical uses of the drug. In many of the medical applications that have been assessed to date, unwanted psychic side effects have been cited as the main reason for patients rejecting the drug as unacceptable. Patients who have had no prior experience with cannabis often find the intoxicant effects disturbing and the drug may induce a frightening panic/anxiety attack in such people. Others may simply not want to be high when they go about their daily work. The deleterious effects of cannabis on short-term memory and other aspects of cognition make it especially unacceptable for those whose occupation depends on an ability to remain alert and capable of handling and processing complex information. If improved delivery systems could be devised it is more likely that patients could self-titrate optimum doses of the drug to avoid some of these unwanted effects, but it appears that the therapeutic window between a medically effective dose and an intoxicant one is narrow.

Along with these psychic effects go impairments in psychomotor skills, so that for a period of some hours after taking the drug it would inadvisable for patients to drive, and their ability to carry out any tasks that require manual dexterity is likely to be impaired. A drug-induced impairment of balance could also make elderly patients more likely to fall. A comparison of 452 marijuana smokers with a similar number of nonsmokers attending the Kaiser Permanente Health Group in California revealed that the marijuana smokers had an increased risk of attending outpatient clinics with injuries of various types — perhaps as a result of the acute intoxicant effects of the drug (Polen et al., 1993).

There are quite profound effects of cannabis on the heart and vascular system. In inexperienced users the drug can cause a large increase in heart rate (up to a doubling) and this could be harmful to someone with a previous history of coronary artery disease or heart failure. Such patients should be excluded from any clinical trials of cannabis-based medicines for this reason. The postural hypotension that can be caused by cannabis could also be distressing or possibly dangerous, as the fall in blood pressure when rising from a seated or a lying down position can result in fainting. The effects of the drug on the cardiovascular system usually show rapid tolerance on repeated exposure to cannabis, so for normal healthy subjects these effects do not appear to be of any particular concern.

Effects of Long-Term Exposure to Cannabis

Introduction

Some alarming claims about the harmful effects of long-term exposure to cannabis were made during the 1970s and 1980s, and many of these continue to be reiterated today:

> Researchers at the University of Mississippi have collected more than 13,000 technical studies on cannabis — hundreds of them pointing to its malign effects.
>
> (D. Moller, *Readers Digest*, September, 1998)

However, as reviewed in Chapters 2 and 3, and by Hollister (1986, 1998) and by Zimmer and Morgan (1997) in their book *Marijuana Myths, Marijuana Facts*, the following myths finally should be put to rest:

1. Cannabis does not cause structural damage to the brains of animals as some reports had claimed, nor is there evidence of long-term damage to the human brain or other than slight residual impairments in cognitive function after drug use is stopped.

2. Although high doses of cannabis or THC can suppress immune system function in animals, there is no evidence of any significant cannabis-induced impairment of immune function in people.

3. High doses of cannabis or THC inhibit the secretion of sex hormones in animals, but there is no evidence that the drug causes any impairment in fertility or sexual function in either men or women.

4. Although there is evidence that cannabis use may be associated with chromosomal abnormalities, the changes are no different from those seen with other widely used drugs (e.g., tobacco and alcohol) and are not present in the germ cells associated with reproduction. The changes seem to be of no clinical significance (Zimmerman and Zimmerman, 1990; Hollister, 1998).

Other possible consequences of the long-term use of cannabis, however, deserve to be examined more closely. As described in the previous chapter, there is a growing recognition that both tolerance and dependence do occur in some chronic users of cannabis. Tolerance to some of the unwanted effects of the drug on the cardiovascular system or to the unpleasant psychic effects may be regarded as positive features, but the possibility of becoming psychologically dependent on the drug is a matter for genuine concern. How important an issue this is in considering the medical use of cannabis remains unclear. Among illicit users of cannabis it seems that only those who regularly consume very large amounts of the drug are at much risk of becoming dependent. The medical users of the drug usually take relatively small doses of cannabis on an intermittent basis and are, therefore, much less likely to become dependent. Case reports from individual patients often stress that they do not want to become high, and that they use the drug only occasionally. In the analogous case of morphine and related opiate pain relievers, medical concerns about the possibility of creating opiate addiction among patients receiving these drugs have been exaggerated. In reality this has not proved to be a serious problem, even though many terminally ill patients receive very large doses of morphine. To quote from the authoritative *Textbook of Pain*:

> The fear of causing psychological dependence is still a potent cause of under-prescription and under-use of strong opioid analgesics. Published data indicate that this fear in unfounded and unnecessary. Among nearly 12000 hospital patients who received strong opioids, there were only four reasonably well-documented cases of addiction in patients who had no history of drug abuse (Porter and Jick, 1980). The dependence was considered major in only one instance, which suggests that the medical use of strong opioids rarely leads to addiction.
>
> (Twycross, 1994)

Cannabis in Pregnancy

There have been warnings that cannabis might cause birth defects ever since the 1960s, and they continue today:

> Some scientific studies have found that babies born to marijuana users were shorter, weighed less, and had smaller head sizes than those born to mothers who did not use the drug. Smaller babies are more likely to develop health problems. Other scientists have found effects of marijuana that resemble the features of fetal alcohol syndrome. There are also research findings that show nervous system problems in children of mothers who smoked marijuana.
>
> Researchers are not certain whether a newborn baby's health problems, if they are caused by marijuana, will continue as the child grows.
>
> (*Marijuana: Facts Parents Need to Know*, National Institute on Drug Abuse, 1997 — http://www.nida.nih.gov/MarjBroch)

A number of studies in animals have shown that THC can cause spontaneous abortions, low birth weight, and physical deformities — but these were only seen after treatment with extremely high doses of THC (50–150 times higher than human doses), and only in rodents and not in monkeys. In chimpanzees, the ape that most closely resembles humans, treatment with high doses of THC for up to 152 days had no effect on fertility or the health of the offspring.

Several studies have compared the babies born to women who had used marijuana during pregnancy with the babies of women who did not. Most studies failed to show any significant differences, but some differences are likely to occur by chance and small differences have been reported in some studies. There is a tendency towards a shorter gestation period and smaller birth weight in babies born to mothers who used marijuana. However, although a significantly lower birth weight was observed in the largest such study (involving 12,424 births), when other factors were taken into account (for example, tobacco smoking) there was no statistically meaningful relation between marijuana use and low birth weight (Zuckerman et al., 1989). Similarly a trend towards a higher incidence of birth abnormalities in the marijuana-exposed group in the same study was also not considered statistically meaningful. If marijuana smoking does cause a reduction in birth weight this is quite likely to be due to the presence of carbon monoxide in marijuana smoke. This gas binds tightly to the red pigment hemoglobin in the blood making it less able to carry oxygen to the growing fetus. It is thought that the carbon monoxide in cigarette smoke is the most likely factor to account for the well-documented effect of tobacco smoking during pregnancy on birth weight.

Several studies have examined the development of children born to mothers who were exposed to marijuana during pregnancy, to see whether any abnormalities in physical or mental development could be detected. While the results of the majority of these investigations were negative, the few instances in which subtle abnormalities could be detected in subsets of the IQ scale have been used as evidence that marijuana can impair children's cognitive development. In one of the largest studies of this kind, psychologist Peter Fried and colleagues examined a group of children whose mothers were exposed to marijuana for the first 6 years of their life. In his *Ottowa Prenatal Prospective Study* hundreds of different psychological tests were administered to the children, but very few differences were found between the marijuana-exposed versus nonexposed groups. The investigators, who appeared convinced that some abnormalities must be there, introduced a new series of cognitive tests when the children were 6 years old, and claimed to have found deficits in frontal lobe executive functions. The differences noted in the babies born to mothers who used marijuana, however, were relatively minor by comparison with the consistent cognitive deficits observed by Fried and colleagues in children of all ages born to mother who had been heavy cigarette smokers during pregnancy (Fried, 1993).

Neither tobacco nor marijuana begins to compare with the serious dangers posed by drinking alcohol during pregnancy. Although this has long been thought undesirable, it was only in the 1970s that fetal alcohol syndrome was first clearly described. Babies born with fetal alcohol syndrome have low birth weight and small heads with abnormal facial features. They have narrow eye slits and a flat face with no groove between nose and upper lip. While their facial features tend to become more normal as they grow older, their mental performance does not improve. They suffer permanent brain damage, which causes quite severe retardation, with IQ scores on an average of 60 or less. Approximately 1 baby in every 1000 in the United States is born with fetal alcohol syndrome, making it the single most important cause of mental retardation. The factors that make some women more susceptible to alcohol than others are not known; consumption of as little as 2 units of alcohol a day can cause fetal alcohol syndrome. There are also much larger numbers of babies born who suffer a lesser degree of damage due to exposure to alcohol during pregnancy, they are described as suffering from "fetal alcohol effect" which can include low birth weight, abnormal facial features, and some degree of mental retardation. As many as 10 in every 1000 babies suffer from such abnormalities.

Returning to cannabis, there are worrying reports that children born to mothers who used marijuana during pregnancy may be liable to a greater risk of developing certain forms of childhood cancer. A small number of studies have reported such an association between marijuana use and rare forms of cancer. The investigators used the so-called "case-control" study design. In this type of research a group of children with cancer and their mothers are compared with an equal number of randomly chosen controls, whose babies were born at about the same time without the disease. One study of 204 children with a form of leukemia (nonlymphoblastic leukemia) found a ten-fold greater incidence of the disease in children born to mothers who had been exposed to marijuana during pregnancy (Robison et al., 1989). This sounds alarming, but if one examines the data more closely, of the 204 mothers whose babies developed leukemia 10 had smoked marijuana during pregnancy, while only 1 of the control group mothers admitted so doing. Of the 10 marijuana smokers only 3 had smoked more than twice a month and only one admitted smoking on a daily basis. There are many other possible differences between the groups, which could explain the result, and no cause and effect relation was established between marijuana and childhood cancer. Another study reported a threefold greater risk for a form of muscle cancer, rhabdomyosarcoma, and similar comments apply (Grufferman et al., 1993). The same authors had previously reported that other risk factors for this cancer included diets that included organ meats, mothers aged over 30, use of antibiotics during pregnancy, and overdue or assisted labor (Grufferman et al., 1982).

Although the risks of exposure to marijuana during pregnancy do not appear to be great, it is surely better not to take any drugs during pregnancy, or to drink alcohol or smoke tobacco.

Cannabis and Mental Illness

The concern that the use of cannabis might precipitate mental illness in some users is a long standing one. There was a lively correspondence in the columns of the *British Medical Journal* in 1893, for example, as to whether or not the endemic use of hashish in Egypt lead to mania and insanity (*Br Med J* 1893, pp 710, 813, 920, 969, 1027). There was also concern that the mental asylums in British India were filling with cannabis-induced lunatics, and this was one of the factors that led the British government to appoint the Indian Hemp Drugs Commission. The commission undertook a large and painstaking review (see Chapter 7) and concluded that there were virtually no patients in the Indian asylums whose illness could be attributed to cannabis use. The commission's findings were not widely noted, however, and claims of a relationship be-

tween cannabis use and insanity continued to be made in India and
many other countries. Claims that cannabis use leads to insanity were
used by early advocates of marijuana prohibition in the United States.

The existence of a temporary form of drug-induced madness in
some cannabis users is, nevertheless, a real phenomenon. In some of the
psychiatric literature, this is referred to as *cannabis psychosis* (or *marijuana psychosis*). Research psychiatrists, particularly in Britain (see
Thomas, 1993 for a review), have carefully studied this condition. It
nearly always results from taking large doses of the drug, often in food or
drink, and the condition may persist for some time, perhaps as the accumulated body load of THC is washed out. The idea that this form of
drug-induced madness is somehow unique to cannabis, however, is probably incorrect. Many powerful psychotropic drugs can precipitate a toxic
psychosis if used incorrectly. With cannabis, as with the other drugs,
patients with a previous history of psychotic illness are most likely to
experience the drug-induced psychosis. Snyder (1971) has pointed out
that the psychoactive effects of cannabis resemble, to some degree, those
elicited by the psychedelic drugs mescaline or LSD. Although the mechanism of action of these drugs differs from that of cannabis, they may
trigger some of the same psychic events in the brain. The acute toxic
psychosis that is sometimes caused by cannabis can be sufficiently serious
as to lead to the subject being admitted to the hospital, and the initial
diagnosis can be confused with schizophrenia, since the patients may
display some of the characteristic symptoms of schizophrenic illness.
These include delusions of control (being under the control of some
outside being or force), grandiose identity, persecution, thought insertion,
auditory hallucinations (hearing sounds, usually nonverbal in nature),
changed perception, and blunting of the emotions. Not all symptoms
will be seen in every patient, but there is a considerable similarity to
paranoid schizophrenia. This has led some to propose a "cannabinoid
hypothesis of schizophrenia," suggesting that the symptoms of schizophrenic illness might be caused by some abnormal over activity of endogenous cannabinoid mechanisms in the brain (Emrich et al., 1997). The
similarity between the drug-induced symptoms and schizophrenic illness
has led to the further suggestion that drugs that block the cannabinoid

CB1 receptor in the brain might prove useful as a treatment for schizophrenia. Clinical trials of the CB1 receptor antagonist SR141716A in schizophrenia are said to be under way.

Another curious phenomenon experienced by some cannabis users is termed a *flashback*. This is a state of altered consciousness resembling a cannabis high that occurs during periods of sobriety after the usual effects of the drug have worn off. The experience can be pleasurable, but more often it is very unpleasant. This can be disconcerting or even dangerous if the person is driving a car or undertaking some other demanding task at the time. The mechanisms involved are not understood, although it is possible that this is merely a bizarre form of déjà vu, or perhaps a sudden acceleration of THC clearance from fat stores could be responsible.

A number of studies during the 1970s and 1980s continued to address the question of a causative link between cannabis and long-term psychiatric illness. The strongest evidence seemed to come from a study in Sweden by Andreasson et al. (1987). The study involved taking detailed medical records, information about the social background, and drug-taking habits of 45,570 conscripts to the Swedish army, and the following up of their subsequent medical history over a 15-year period. A total of 4293 of the conscripts admitted having taken cannabis at least once, but the cannabis users accounted for a disproportionate number of the 246 cases of schizophrenic illness diagnosed in the overall group on follow-up. The relative risk of schizophrenia in those who had used cannabis was 2.4 times greater than in the nonusers. And in the small number of heavy users (who had taken the drug on more than 50 occasions) the relative risk of schizophrenia increased to 6.0. The authors concluded that cannabis was an independent risk factor for schizophrenia. At first sight these findings seem convincing, but the authors' conclusions have been widely criticized. It is notable that the cannabis-taking group also admitted to using a variety of other psychoactive drugs, and the findings do not prove any cause-and-effect relationship with cannabis. It may be simply that both cannabis use and schizophrenia may be related to some common predisposing factor such as personality. Indeed some psychologists and psychiatrists believe that they can identify psychological traits that are described as *schizotypy* and that may predict an increased risk of developing clinical psychosis. One study in 211 healthy adults found that those subjects who used cannabis scored higher on schizotypy scales than nonusers (Williams et al., 1996). Andreasson et al. (1989) in a more detailed follow-up of some of the original cohort claimed to have answered some of these criticisms, but their results were far from conclusive. More than half of the cannabis-using subjects who developed schizophrenia had also taken amphetamine — a drug known to be capable of inducing a schizophrenia-like psychosis. The cannabis users also came from deprived social backgrounds, another known risk factor of schizophrenia.

The existence of any causative relationship between cannabis use and long-term psychotic illness thus seems unlikely to most people. If cannabis use did precipitate schizophrenia one might expect to have seen a large increase in the numbers of sufferers from this illness as cannabis use became more common in the West during the past 30 years. A detailed review of the epidemiological evidence by Thornicroft (1990), however, showed that this has not been the case.

It seems likely, however, that cannabis can exacerbate the symptoms of existing psychotic illness. In patients suffering from schizophrenic illness, cannabis made the key symptoms of delusions and hallucinations worse and tended to counteract the antipsychotic effects of the drugs used to treat the illness (Negrete et al., 1986; Linzen et al., 1994). On the other hand, one Swedish study reported that cannabis use made schizophrenic patients less withdrawn and more likely to speak (Peralta and Cuesta, 1992). It would seem prudent, nevertheless, to discourage the use of cannabis in patients with existing psychotic illness.

Special Hazards of Smoked Marijuana

Traditionally, the use of cannabis both in Oriental and Western medicine involved taking the drug by mouth, but most of the current illicit medical use of the drug in the West involves the inhalation of marijuana smoke. Smoking is a remarkably efficient means of delivering an accurately gauged dose of THC but unfortunately, it carries with it special hazards. Although THC itself appears to be a relatively safe drug, the same cannot be said of marijuana smoke.

Marijuana Smoke and Smoking Behavior

Although relatively little research has been done on the effects of marijuana smoke, a great deal is known about the toxic components in tobacco smoke and their biological effects. Marijuana smoke is very similar in chemical composition to tobacco smoke, so it is not unreasonable to suggest that our knowledge of the dangers of tobacco can provide useful predictions about the hazards of smoked marijuana. A burning tobacco cigarette has been described as a "miniature chemical factory." In addition to the large number of chemical components present in the dried plant material, hundreds of additional chemicals are created during the process of combustion. More than 6000 chemical constituents have been identified in tobacco smoke and thousands more are present in trace amounts. The composition of tobacco smoke varies according to the manner in which the material is smoked. The nature of the wrapping paper, for example, alters the burning characteristics and consequently alters the chemical composition of the smoke. There is no reason to think that the same considerations do not also apply to marijuana. Table 5.1 summarizes the components present in a typical cigarette or marijuana joint. Apart from the fact that the former contains nicotine whereas the latter contains THC, the profiles are otherwise remarkably similar. Smoke consists of two components, the minute droplets present in the particulate phase, and the various volatile chemicals or gases in the vapor phase. About 10% of the total weight of fresh tobacco or marijuana smoke is in the particulate phase, which contains most of the active drug (nicotine or THC). The particulate phase consists of minute droplets of condensed fluid, less than a quarter of a millionth of a meter in diameter (less than 1/1000 of a millimeter), with as many as 5 billion droplets per milliliter of smoke. Both vapor and particulate phases of both marijuana and tobacco smoke contain a number of toxic chemicals, several of which are known to be capable of promoting the development of

Table 5.1. Composition of Mainstream Smoke from Marijuana and Tobacco Cigarettes

	Marijuana Cigarette	Tobacco Cigarette
Average weight (mg)	1115.0	1110.0
Moisture (%)	10.3	11.3
Gas Phase		
Carbon monoxide (mg)	17.6	20.2
Carbon dioxide (mg)	57.3	65.0
Ammonia (mg)	0.3	0.2
Hydrogen cyanide (μg)	532.0	498.0
Cyanogen (μg)	19.0	20.0
Isoprene (μg)	83.0	310.0
Acetaldehyde (μg)	1200.0	980.0
Acetone (μg)	443.0	578.0
Acrolein (μg)	92.0	85.0
Acetonitrile (μg)	132.0	123.0
Benzene (μg)	76.0	67.0
Toluene (μg)	112.0	108.0
Vinyl chloride (ng)*	5.4	12.4
Dimethylnitrosamine (ng)*	75.0	84.0
Methylethylnitrosamine (ng)*	27.0	30.0
Particulate Phase		
Total particulate matter (mg)	22.7	39.0
Phenol (μg)	76.8	138.5
o-Cresol (μg)	17.9	24.0
m- & p-Cresol (μg)	54.4	65.0
Dimethylphenol (μg)	6.8	14.4
Catechol (μg)	188.0	328.0
Cannabidiol (μg)	190.0	—
Δ^9-THC (μg)	820.0	—
Cannabinol (μg)	400.0	—
Nicotine (μg)	—	2850.0
N-nitrosonornicotine (ng)*	—	390.0
Naphthalene (μg)	3.0	1.2

(*continued*)

Table 5.1. —Continued

	Marijuana Cigarette	Tobacco Cigarette
1-methylnaphthalene (μg)	3.6	1.4
Benz(a)anthracene (ng)*	75.0	43.0
Benz(a)pyrene (ng)*	31.0	21.1

*Indicates known carcinogens.

[Data from British Medical Association Report 1997]

cancers (carcinogens). Some reports have indicated that two of the most potent known carcinogens in tobacco smoke, benzanthracene and benzpyrene are present in even higher amounts in marijuana smoke, although other measurements indicate that the amounts are similar in both types of smoke.

The way in which experienced users smoke marijuana tends to enhance the potential dangers of taking the drug by this route. Marijuana smokers usually inhale more deeply than tobacco smokers and they tend to hold their breath, in the belief that this increases the absorption of THC by the lungs. (In fact, the results of experimental studies in which both puff volume and breath-hold duration were systematically varied show that while inhaling more deeply does increase the amount of THC absorbed, holding the breath for more than a few seconds has rather little effect. The concept seems to be based more on cultural myths than on reality). The results of these differences in smoking behavior are quite profound. Wu et al. (1988) compared the amounts of particulate matter (tar) and carbon monoxide absorbed in 15 volunteers who were regular tobacco and marijuana smokers. Results were compared after smoking a single filter-tipped tobacco cigarette or a marijuana cigarette of comparable size. As compared with smoking tobacco, smoking marijuana resulted in a fivefold greater absorption of carbon monoxide and 4–5 times more tar was retained in the lungs (Table 5.2).

A number of studies have explored how marijuana smoking behavior varies according to how much THC is present in the cigarette. Not surprisingly, experienced marijuana smokers automatically alter their smoking behavior to achieve the desired high, regardless of whether they are told that the cigarette is of high or low potency. When smoking high

Table 5.2. Comparison of Smoking Marijuana Versus Tobacco Cigarette

	Tobacco	Marijuana
Puff volume (ml)	49.4 ± 15.2	78.0 ± 22.8
Puff duration (seconds)	2.4 ± 1.1	4.0 ± 2.2
No. of puffs	13.5 ± 4.0	8.5 ± 3.1
Interval between puffs (seconds)	27.0 ± 8.2	37.6 ± 14.5
Inhaled volume (liters)	1.31 ± 0.22	1.75 ± 0.52
Smoke-retention time (sec)	3.5 ± 1.3	14.7 ± 10.2
Inhaled particulates (O.D.)	4.9 ± 2.0	16.3 ± 6.3
Particulates deposited (%)	64.0 ± 8.9	86.1 ± 6.71

Data are averages with 95% confidence limits obtained from 15 volunteers. Inhaled particulates were assessed by optical density (O.D.) measurements. From Wu et al. (1988).

potency cigarettes the total volume inhaled was less and there was significantly less tar deposited. The conclusion seems to be that habitual marijuana smokers could reduce the health hazards of smoking by using
marijuana with a high THC content. Other possibilities include the development of strains of cannabis plants that produced a lower yield of tar,
or the use of filters or other devices to reduce the tar content of marijuana smoke before it enters the lungs. Neither of these alternatives
seems to have been investigated in any systematic way to date.

Effects of Marijuana Smoke on the Lungs

Since tobacco smoking is known to be the most important cause of
chronic obstructive lung disease and lung cancer, it is reasonable to be
concerned about the adverse effects of marijuana smoke on the lungs.
There have been a number of attempts to address this question by expos-
ing laboratory animals to marijuana smoke. After such exposure on a
daily basis for periods of up to 30 months, extensive damage has been
observed in the lungs of rats, dogs, and monkeys, but it is very difficult to
extrapolate these findings to man as it is difficult or impossible to imitate
the human exposure to marijuana smoke in any animal model. The various studies that have been undertaken in human marijuana smokers
seem far more relevant, although here the problem is confounded by the

fact that many marijuana smokers consume the drug with tobacco, making it difficult to disentangle the effects of the two agents. Professor Donald Tashkin and his colleagues at the Department of Medicine at the University of California Los Angeles have been a leaders in this field for more than a decade (for review see Tashkin, 1999). Although a number of studies in the 1970s had reported an association between marijuana smoking and chronic bronchitis, the number of subjects examined was small and there was a lack of control for the important confounding variable of tobacco smoking. In 1987, Tashkin reported the results of the first large scale study of 144 volunteers who were heavy smokers of marijuana only. He compared these with 135 people who smoked tobacco and marijuana, as well as 70 smokers of tobacco only and 97 non-smokers. Approximately 20% of both tobacco smokers and marijuana smokers reported the symptoms of chronic bronchitis (chronic cough and phlegm production), even though the marijuana smokers consumed only 3–4 joints a day versus > 20 cigarettes for the tobacco smokers. In this study no additive effects were seen in those who smoked both marijuana and tobacco, although additive effects have been reported in other studies of this type. Ten years later Tashkin described a follow-up study of the groups studied earlier. He found that lung function in the tobacco smokers had continued to get worse over the 10-year period, particularly in the small airways, making them more liable to develop chronic obstructive lung disease later in life. No such decline was observed, however, in the marijuana smokers, suggesting that they may be less likely to develop such diseases as emphysema because of their smoking. Similar conclusions were reached from a study of 268 heavy marijuana smokers in Australia. After smoking regularly for an average of 19 years they had a lower prevalence of asthma or emphysema than the general population. At the Kaiser Permanente Health Care group in California, a careful comparison of 452 daily marijuana smokers who never smoked tobacco with 450 nonsmokers of either substance revealed that the marijuana smokers had a small increased risk of outpatient visits for respiratory illness (Relative risk = 1.19) (Polen et al., 1993).

Some of the volunteers from Tashkin's Los Angeles study were subjected to a saline rinse of their lungs in order to sample the population of white blood cells present. White cells are the soldiers of the immune system, they are attracted to regions of tissue inflammation or damage and help to kill and remove infectious microbial invaders and to remove damaged or dead cells and tissue debris. The large white cells known as macrophages are particularly important scavengers, which engulf and kill invading bacteria and fungi and remove damaged tissue. Approximately 2–3 times more macrophages were collected from the lungs of tobacco or marijuana smokers versus nonsmokers, suggesting the presence of an inflammatory response. The macrophage from both tobacco and mari- juana smokers also showed significant impairments in their ability to kill and engulf fungi (*Candida albicans*) or bacteria (*Staphylococcus aureus*). The macrophages from smokers were also less able to generate some of the chemical toxins (e.g., superoxide) that they normally use to kill in- vading microorganisms, or the chemicals known as cytokines that help to activate further inflammatory and immune system responses. In addition, the macrophages were impaired in their ability to attack and kill cancer cells (small cell cancers) in vitro. Studies in animals have confirmed these findings, showing that exposure of macrophages to marijuana smoke in vitro impairs their function, and that exposure of rats to mari- juana smoke in vivo makes them less able to inactivate bacteria (*Sta- phylococcus aureus*) delivered by aerosol to the lungs. The animal studies also indicated that the toxic effects of marijuana smoke on the immune defenses were not due to THC but to some other components of the marijuana smoke, since smoke from THC-extracted herbal material re- mained toxic. These findings suggest that like tobacco smokers, mari- juana smokers are likely to be more susceptible to respiratory tract infec- tions and possibly less able to defend against the development of lung cancers. An added complication is that some batches of herbal cannabis may be contaminated with fungi (e.g., *Aspergillus* species) that could themselves cause lung infections. This could be a particular hazard to AIDS patients whose immune defenses are already compromised.

A similar concern is the contamination of some United States sup- plies of herbal cannabis by the herbicide paraquat, used by the United States government to destroy cannabis crops in the United States and

Mexico. Paraquat has potent toxic effects on the lung, causing inflammation and congestion that can be life threatening. Fortunately this hazard seems less important now than it was a few years ago.

Marijuana Smoking and Lung Cancer

Tetrahydrocannabinol does not appear to be carcinogenic, but there is plenty of evidence that the tar derived from marijuana smoke is. Bacteria exposed to marijuana tar develop mutations in the standard Ames test for carcinogenicity and hamster lung cells in tissue culture develop accelerated malignant transformations within 3–6 months exposure to tobacco or marijuana smoke. Painting marijuana tar on the skin of mice also leads to premalignant lesions. But is there any evidence that this happens in the lungs of marijuana smokers?

As part of Tashkin's original 1987 study, some of the volunteers were examined in more detail for evidence of damage to the airways. Visual examination of the large airways with a bronchoscope showed that a large proportion of both marijuana and tobacco smokers showed evidence of increased redness and swelling and increased mucus production relative to nonsmokers. Excision of minute amounts of tissue (biopsies) from the lining of the airways allowed microscopic examination. This revealed abnormal cell changes in both marijuana and tobacco smokers. These included an abnormal proliferation of mucus-producing cells and a reduced number of ciliated cells (these are normally present in the lining of the airways; the movement of their hair-like cilia helps to clear the lungs of mucus and debris). These changes could explain the chronic cough and overproduction of phlegm reported by tobacco and marijuana smokers. A more sinister observation was the presence of abnormal cells resembling those normally seen in skin (squamous metaplasia) in the lungs of smokers. These changes are thought to represent premalignant precursors for the development of lung cancer. The possible premalignant cells were seen to an even greater extent in the lungs of volunteers who smoked both marijuana and tobacco. Tashkin and his colleagues have extended these studies more recently by examining lung biopsies from marijuana and tobacco smokers to see if the cells express certain genes that must be activated for normal lung cells to transform into cancer cells. This is a complex process that involves the switching on of various genes that control cell growth and proliferation. Evidence was found for an over expression of genes controlling receptors for epidermal growth factor and a nuclear proliferation protein responsible for cell division known as KI-67 in the lungs of smokers. Tashkin's group also reported that bronchial biopsies from marijuana smokers showed evidence of over expression of the enzyme CYP1A1. This is a member of the cytochrome p450 family of enzymes responsible for drug metabolism in the liver and other tissues. The significance of the CYP1A1 enzyme is that it is known to play a key role in converting the benz[a]pyrene present in tobacco smoke into a very potent carcinogen. The p53 oncogene, which is known to play a role in 75% of lung cancers, was not activated, however, except in a single subject who was a combined marijuana and tobacco smoker. The changes observed are worrying as they may indicate that a series of precancerous changes take place in the lungs of marijuana smokers, similar to those that occur in tobacco smokers, the end result of which may be to significantly increase the likelihood of developing lung cancer.

The discovery of the link between cigarette smoking and lung cancer was one of the great achievements of medical research in this century. The initial reports in 1950 from Britain and the United States were based on two very large case-control studies. Subsequently a great deal more has been learned from follow-up studies in large groups of smokers and nonsmokers. One such study involved asking all the doctors in Britain about their smoking habits. More than 40,000 doctors agreed to take part in a long-term study to see what effect their smoking habits might have on their health. The study started in 1951, and Doll et al. (1994) described the results of a 40-year follow-up of this group. The results are alarming; not only was the risk of dying from lung cancer increased in the cigarette smokers, but so were the risks of dying from 23 other causes, including cancers of the mouth, throat, larynx, pancreas, and bladder and such obstructive lung diseases as asthma and emphysema. The authors concluded that:

Results from the first 20 years of this study, and of other studies at that time, substantially underestimated the hazards of long term use of tobacco. It now seems that about half of all regular cigarette smokers will eventually be killed by their habit.

It is hard to overestimate the importance of tobacco smoking as the principal avoidable cause of death in the modern world. More people die from smoking tobacco than any other single cause. Worldwide some 3 million deaths a year can be attributed to tobacco, and this is likely to rise to 10 million a year in 30–40 years' time (Peto et al., 1996). In developed countries tobacco is responsible for nearly a quarter of all male deaths and 17% of women. People in the developing countries started smoking later in the twentieth century, but they are catching up fast in the tobacco mortality statistics. The results of a recent study in China, involving an analysis of more than 1 million deaths, makes some frightening predictions. Cigarette smoking in China has increased dramatically in the recent past—almost quadrupling since 1980. About two thirds of men over the age of 25 smoke and about half of these will die prematurely. This implies that eventually 100 million of the 300 million young men now alive and aged 0–29 will be killed by tobacco (half dying in middle age, half in old age) (Liu et al., 1999).

One of the reasons why we should be seriously concerned about the possible link between marijuana smoking and lung cancer is that it could take a very long time for such a relationship to become manifest. Cigarette smoking became common among men in the developed world during the first decades of this century, but it was not until 30–40 years later that the first evidence of a link between tobacco smoking and lung cancer was obtained. Even though cigarette consumption has declined significantly in many developed countries, deaths from tobacco-related diseases will continue to rise for many years to come, particularly among women for whom cigarette smoking was not common until the 1930s or 1940s. Such long lag periods between cause and effect are hard to comprehend. The relationship between cigarette smoking and lung cancer is very complex. The increased risk of developing lung cancer depends far more strongly on the duration of cigarette smoking than on the number of cigarettes consumed each day. Thus, while smoking three times as many cigarettes a day does increase the lung cancer risk approximately threefold, smoking for 30 years as opposed to smoking for 15 years does not simply double the lung cancer risk, it increases the risk by 20-fold, and smoking for 45 years as opposed to 15 years increases the lung cancer risk 100-fold (Peto, 1986).

The reasons underlying the relationship between the duration of tobacco smoking and the development of lung cancer are unknown, but they are quite likely to apply to marijuana smokers as well. To argue as some have done (Zimmer and Morgan, 1997), that because a link has not yet been established between marijuana smoking and lung cancer and therefore no such link is likely to exist, is meaningless. Since the widespread use of marijuana as a recreational drug is a fairly recent development in the Western world, large numbers of people have not yet been exposed to marijuana smoking for long enough for any link to become clear. The following comments on tobacco smoking could well apply also to marijuana:

> Among regular cigarette smokers, the excess lung cancer risk depends strongly not only on smoking habits during the past few years, but also on smoking habits during early adult life. Hence, current lung cancer rates in countries where smoking among young adults became widespread less than half a century ago may be serious underestimates of the eventual magnitude of the tobacco-induced lung cancer hazard.
>
> (Peto, 1986)

One of the few large-scale studies of the health consequences of marijuana smoking was reported by Sidney et al. (1997). The authors studied a cohort of 65,171 men and women undergoing health checks at the Kaiser Permanente Health Group in California between 1979 and 1985. The health of these subjects was then followed for an average of 10 years. A total of nearly 27,000 people admitted to being either current or former marijuana users (defined as ever having smoked more than six times). Over the period of the study 182 tobacco-related cancers were detected, of which 97 were lung malignancies. No effects of former or current marijuana use on the risk of any cancers were found. However, although this study involved large numbers, almost all the marijuana smokers were young (15–39) and the follow-up period was relatively short. Such a study could not have been expected to detect any relationship between marijuana and lung cancer if the lag period were comparable to that seen with tobacco. It may not be possible to answer the question of a link between marijuana smoking and lung cancer for another decade or more. Meanwhile we must hope that young people in the West are not storing up a time bomb that may shorten their lives as tobacco smoking has done to earlier generations.

There are some mitigating features of this otherwise somber topic. One of the first groups in our society who understood the message about the dangers of cigarette smoking were doctors. In the United States and in Europe doctors were among the first to change their smoking habits as a consequence of their appreciation of its dangers. Consequently the follow-up study of British doctors already referred to contained a substantial number of doctors who gave up smoking. Although the added risk of lung cancer that cigarette smoking confers does not go away, the risk becomes so much greater with increased duration of exposure that those who give up smoking benefit disproportionately. In fact, the British doctors who gave up smoking before the age of 35 had a pattern of survival that did not differ significantly from nonsmokers (Fig. 5.1). Those who gave up smoking when older had a survival rate that was between that of continuing smokers and nonsmokers. The relevance of this to marijuana smokers is clear. A number of surveys have indicated that the majority of marijuana smokers are in their teens or twenties and they tend to give up the habit when they reach their thirties. If the pattern is similar to that seen with tobacco smoking, then their risk of developing lung cancer later in life may not be significantly increased. On the other hand, some surveys of current marijuana use have indicated that more users are continuing to smoke throughout their lives. Therefore, the previous pattern of quitting in midlife may no longer be true in the future.

Another factor to consider is how much tar the marijuana smoker is exposed to vis à vis the cigarette smoker. Although the marijuana smoker on average consumes no more than 3–4 joints a day—in contrast to the

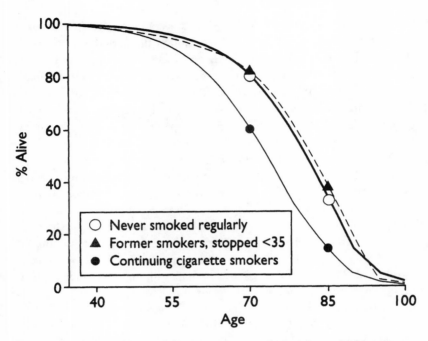

Figure 5.1. Effects on survival of stopping smoking tobacco before the age of 35. Results from a survey of about 40,000 British doctors; those who continued smoking had their life expectancy reduced by about 8 years, but those who gave up smoking before age 35 had a life expectancy not significantly different from that of nonsmokers. From Doll et al. (1994). Reprinted with permission from BMJ Publishing Group.

15–20 cigarettes commonly consumed by tobacco users — each joint is liable to deposit 4–5 times more tar in the lungs than a tobacco cigarette. The tar exposure is thus similar, except for the large numbers who smoke marijuana with tobacco; for them the hazards are compounded.

At the moment the jury is out on the link between marijuana and cancer, although there has been some concern about reports of an increased number of cancers of the tongue and larynx in young people with a history of heavy marijuana use (see Tashkin, 1999). These are based, however, on very small numbers and no cause and effect relationship has been established. One report, for example, examined the pathology records of a hospital in Florida and identified 10 patients under the age of 40 out of a total of 887 who developed cancer of the respiratory tract. Of the 10 young patients, 7 had a history of moderate to heavy marijuana use, 1 was a "probable" user, and 2 had no known history of marijuana use.

Can Smoked Marijuana Be Recommended for Medical Use? Are There Alternative Delivery Systems?

Given the well documented adverse effects of smoked marijuana on the lungs and the possible link to cancers of the upper and lower respiratory tract is there any place at all for smoked marijuana in medicine? Apart from the potential respiratory hazards, the idea of a smoked herbal remedy goes against the grain of much of our thinking in scientifically based medicine. As the American Medical Association (1997) put it:

> . . the concept of burning and inhaling the combustion products of a dried plant product containing dozens of toxic and carcinogenic chemicals as a therapeutic agent represents a significant departure from the standard drug approval process. According to this viewpoint, legitimate therapeutic agents are comprised of a purified substance(s) that can be manufactured and tested in a reproducible manner.

On the other hand, there is little doubt that for many patients smoking provides a superior method of delivering THC than taking THC or cannabis extracts by mouth. Because of the variable and delayed absorption of orally administered THC the patient is always exposed to the possibility of either under- or overdosing. Smoking, on the other hand, with some practice, permits the rapid delivery of what the individual patient judges to be the correct therapeutic dose. It is clear that more research is urgently needed on alternative methods for rapidly delivering precisely gauged doses of THC, and this has been a recommendation given some priority in several recent official reports (American Medical Association 1997; US National Institutes of Health, 1997; British Medical Association, 1997; House of Lords, Science and Technology Committee, UK, 1998; Institute of Medicine, USA 1999). There have been attempts to deliver THC as an inhaled aerosol, but this seems to cause unacceptable irritant effects on the respiratory system. Perhaps some nonirritant means of delivering the drug to the lung, or to the nasal cavity could be devised. The lung, and to a lesser extent the nasal cavity, have large surface areas with a rich blood supply that represent attractive routes for drug delivery in general — not just for the local treatment of lung or airway conditions. Another concept is to devise a vaporizer that would heat herbal cannabis or pure THC to near the temperature at which the drug vaporizes without igniting it, so that the user could inhale the active drug without the dangerous products of combustion. A number of such devices have been devised by the marijuana smoking community and some are advertised on web sites. There are few reports of scientific studies of their effectiveness although some reports suggest that currently available vaporizers still produce an unacceptable level of tar relative to THC in the inhaled smoke. The same studies also showed that water pipes are not very effective in reducing the inhaled tar to THC ratio, and that adding standard cigarette filters to marijuana cigarettes may make matters worse rather than better as these may be more effective in removing THC than tar (House of Lords 1998, Scientific Evidence. P235). Research on alternative delivery systems for THC has recently been taken up more seriously by some pharmaceutical companies so we may see real advances in this field in the future.

Meanwhile what place, if any, should smoked marijuana have in modern medicine? Because of the potential hazards of chronic respiratory disease and cancer it is unlikely that smoked marijuana could ever be recommended for the long-term treatment of any illness where its use might need to be continued on a regular basis for many years. But if one considers the principal groups currently using smoked marijuana these consist mainly of patients with serious life-threatening illnesses. Patients suffering from AIDS, cancer, or multiple sclerosis have a considerably reduced life expectancy because of their illness — it could plausibly be argued that the long-term health risks of smoking marijuana are of little relevance to such patients. If their illness does not respond to conventional medicine, and their doctor has agreed that smoked marijuana might benefit them, why should the law stand in their way? The House of Lords Science and Technology Committee (1998) were persuaded by this concept of the "compassionate reefer" and recommended that cannabis be rescheduled (see Chapter 6) so as to allow doctors to prescribe it on a named patient basis. As their report puts it:

> Our principal reason for recommending that the law be changed, to make legal the use of cannabis for medical purposes is compassionate. Illegal medical use of cannabis is quite widespread; it is sometimes connived at and even in some cases encouraged by health professionals; and yet it exposes patients and in some cases their carers to all the distress of criminal proceedings, and the possibility of serious penalties.

Voters in a number of states in the United States share this view and have voted in favor of proposals to make cannabis, including smoked marijuana, available for medical use (see Chapter 7). Despite its hard line on the unscientific nature of smoked marijuana as a medicine, the American Medical Association report (1997) nevertheless recommended that:

> . . . adequate and well-controlled studies of smoked marijuana be conducted in patients who have serious conditions for which preclinical, anecdotal, or controlled evidence suggests possible efficacy including AIDS wasting syndrome, severe acute or delayed emesis induced by chemotherapy, multiple sclerosis, spinal cord injury, dystonia and neuropathic pain. . . .

The report of an NIH expert group published in the same year also recommended that controlled studies be done to compare the efficacy of smoked marijuana with orally administered dronabinol in various potential target illnesses. The influential Institute of Medicine (1999) report also concluded that although smoked marijuana should generally not be recommended for long-term use, there were certain patients for whom short-term use of the smoked drug could be justified.

Comparison of the Health Risks of Cannabis with Those of Alcohol and Nicotine

It seems obvious to ask the question, how do the health risks of cannabis compare with those attributable to the other two most commonly used psychotropic drugs, alcohol and nicotine? This comparison, however, is fraught with difficulties. Alcohol and tobacco are used by far larger numbers of people than cannabis, so their impact on public health is correspondingly greater. We also know so much more about the long-term effects of tobacco and alcohol on health than we do about cannabis, consequently any comparison almost inevitably makes cannabis appear to be the safer drug. This in turn leaves the author open to the accusation that the arguments are being rehearsed in order to promote the relaxation of current prohibitions on cannabis use. Such arguments lead a recent review comparing the health and psychological effects of cannabis, alcohol, nicotine, and opiates to be withdrawn at the last moment from the published WHO Report (1997). Nevertheless Hall et al. (1994) in their excellent review of *The Health and Psychological Consequences of Cannabis Use* attempted such a comparison. It is worth reminding ourselves of the well-documented health risks that we are willing to tolerate in recreational drugs.

Alcohol

The major acute risks of alcohol are similar to those of cannabis, namely those associated with intoxication. Both drugs cause impairments in psychomotor and cognitive function, especially memory and forward planning. Alcohol intoxication increases the risks of being involved in road traffic and other accidents, as cannabis probably does, but unlike cannabis alcohol tends to encourage aggressive behavior. It is an important factor in domestic violence. Unlike cannabis, acute alcohol intoxication can also be life threatening.

There are a number of health risks associated with chronic alcohol intake. These include damage to the unborn child (fetal alcohol syndrome), increased risk of cancers of the mouth and throat, serious damage and cirrhosis of the liver, and permanent damage to the brain (Korsakoff's syndrome) leading to severe cognitive impairments. Alcohol also causes dependence in some people and this is associated with potentially life-threatening withdrawal symptoms when alcohol use is stopped. The heavy use of alcohol can lead to serious impairment of work performance and family life, and dependent users may become psychotic. Cannabis can also cause cognitive impairments and some users may become dependent but the withdrawal symptoms are relatively mild. In the United States more than 150,000 deaths each year are attributable to alcohol abuse.

Tobacco

The acute effects of smoking tobacco are similar to those caused by smoking marijuana, namely the irritant effects of smoke on the respiratory system and the stimulant effects of nicotine or THC on the cardiovascular system. Chronic use in each case leads to an increased risk of developing bronchitis, and in the case of tobacco use, other such respiratory diseases as emphysema and asthma. Tobacco smoking poses a major public health risk because of the established link with lung cancer and several other forms of cancer. It is possible that cannabis smoking also carries with it increased risks of developing cancers of the lungs and upper regions of the airways, but these are not yet proven. In the United States some 350,000 deaths each year are due to tobacco-related diseases.

Summary of the Conclusions Regarding the Medical Uses of Cannabis

1. The only medical uses for which there is rigorous scientific evidence are in the treatment of the sickness associated with cancer chemotherapy and to counteract the loss of appetite and the wasting syndrome in AIDS. There is, however, scientific evidence to support the potential use of cannabis in several other medical conditions, particularly those associated with painful muscular spasms and possibly other forms of treatment-resistant clinical pain. Only anecdotal evidence is available so far in such conditions as multiple sclerosis, spasticity, spinal cord injury, migraine, glaucoma, or epilepsy.

2. The safety profile of THC, the active ingredient of cannabis, is good. It has very low toxicity both in the short and in the long term.

Some of the acute effects of the drug, however, including unpleasant psychic reactions, intoxication, and temporary impairments in skilled motor and cognitive functions, limit the usefulness of THC as a medicine. There appears to be only a narrow window between the desired and the undesired effects.

3. Because of the cardiovascular effects of THC and its propensity to make schizophrenic symptoms worse, patients with cardiovascular disease or schizophrenia are not suitable subjects for cannabis-based medicines. As with most other CNS drugs, cannabis use should be avoided during pregnancy.

4. The safety of smoked marijuana is far more questionable. It causes chronic bronchitis in a substantial proportion of regular users, and the risk that in the longer term an association may be found with cancers of the respiratory tract makes it unsafe to recommend for any long-term use. The compassionate use of smoked marijuana in certain categories of severely ill patients may, nevertheless, be justified.

5. In all instances, including the use of smoked marijuana, more properly controlled clinical trials are needed and research on improved means of delivering the drug is also of high priority.

The Institute of Medicine (1999) in their report *Marijuana and Medicine* summarized the safety issues succinctly:

> Marijuana is not a completely benign substance. It is a powerful drug with a variety of effects. However, except for the harms associated with smoking, the adverse effects of marijuana use are within the range of effects tolerated for other medications.

6

The Recreational Use of Cannabis

The use of cannabis as a recreational drug was almost unknown in the West until the 1950s and only became widespread during the 1960s. The exposure of large numbers of young American soldiers to cannabis during the Vietnam War was an important contributory factor (see for example the famous Vietnam War movies *Platoon* and *The Deer Hunter*). As Napoleon's army brought cannabis to Europe from their Egyptian campaign, the returning soldiers brought cannabis to the United States. The use of cannabis by young people on both sides of the Atlantic was closely linked to the protest and rebellion experienced by the 1960s generation:

> The most profound example of the ability of marijuana to raise mass social consciousness occurred during the Vietnam War era, on both the home front and the battle front. The spread of marijuana use to almost an entire generation of middle-class youth who came of age in the 1960's is inextricable from the dramatic changes in social, political, spiritual and cultural values that mark that era. Cannabis did not kidnap them or their collective consciousness: the generation was ready for marijuana."
>
> (Robinson, 1996).

By the end of the twentieth century another generation has replaced the rebellious youth of thirty years ago; a new generation far less extravagant in their lifestyle, more serious, and no longer feeling the deep sense of alienation from traditional society that many young people experienced in the 1960s. To this generation cannabis is a part of their culture, no longer a gesture of rebellion. Many of the parents of today's generation of cannabis users themselves belonged to the 1960s and 1970s group of marijuana smokers.

Cannabis has become by far the most widely used illicit drug in the West. It ranks as the third most commonly used recreational drug after alcohol and tobacco. Whereas detailed information is available on the consumption of alcohol and tobacco, the health problems they cause and the consequent economic costs to society, such information is largely lacking for cannabis. The use of cannabis occurs in an underground world of illegality. In most countries, according to a United Nations Convention, cannabis is considered a Schedule 1 drug i.e., a dangerous narcotic with no accepted medical usefulness. Possession of cannabis, cultivation of the cannabis plant, or the trafficking of cannabis are all criminal offenses, some of which can carry severe penalties. It is not surprising that the users and suppliers of this illicit drug are not always willing to provide detailed information about it.

Prevalence

For reasons mentioned above it is difficult to obtain accurate figures on the prevalence of cannabis use, but there are some useful sources. A WHO Report titled *Cannabis: a Health Perspective and Research Agenda* published in 1997 provided a summary of data from several countries around the world (WHO, 1997). In the United States the National Household Survey on Drug Use has produced a valuable annual report since 1972; in the United Kingdom the British Crime Survey includes data on drug use. As many as one-third of the entire population aged 15–50 in many Western countries admit to having used cannabis at least once (Table 6.1) Consumption is highest in the younger age groups, so

Table 6.1. Lifetime Prevalence of Cannabis Use Around the World (15–50 Year Olds)

Country and Year of Survey Data	% Ever Used Cannabis
Denmark (1994)	37
Australia (1993)	34
United States (1996)	32
Jamaica (1994)	29
United Kingdom (1994)	20
Switzerland (1991)	17
Germany (West) (1994)	14
Peru (1994)	8
India (South) (1991)	7
Guatemala (1994)	7
Colombia (1994)	6

(WHO Report, 1997)

for 18-year-olds in the United States and many European countries this figure rises to almost 50%.

Patterns of consumption over the years have varied differently in various countries. In the United States, statistics provided by the National Institute on Drug Abuse through their *Monitoring the Future Study* give a detailed picture of cannabis use among teenagers. Cannabis became very popular among young people in the United States during the 1970s, reaching a peak in 1979 when more than 60% of 12th grade students in American high schools (average age 18) admitted ever having used the drug, and 10% reported that they were daily users. There was then a marked drop in consumption during the 1980s. Consumption has, however, increased again rapidly during the1990s, and among teenagers it now approaches the 1979 levels (Fig. 6.1). The results of a study published in 1997 gave data on a postal survey of 17,592 students at 140 American colleges (Bell et al., 1997). One in four of the students who responded (24.8%) reported using marijuana during the past year. The 1990s saw a particularly sharp increase in cannabis use by younger people, from 10% in 1991 to 23% in 1997 among American 8th grade students (average age 14) who admitted to ever having used the drug. Patterns of consumption in most European countries have lagged somewhat behind those in the United States but most countries did not experience the substantial drop in consumption seen in the United States during the 1980s. Current levels of consumption in Europe are now similar to those in the United States.

The great majority of people who try cannabis do so experimentally. Unlike tobacco, where a high proportion of first time users go on to become lifetime smokers, most cannabis users do not go on to become regular users of the drug. Thus, 1996 data in the United States show that while 68.6 million people (32%) admitted having used cannabis once, only 18.4 million (8.6%) reported having used the drug during the past year. In Britain in 1994, 20% admitted ever having used cannabis, but only 5% had used the drug during the past year. It is difficult to assess how many people are regular users of cannabis. There is no agreed definition of "regular user," it could mean anything from someone who took the drug a few times a year on special occasions, to someone consuming

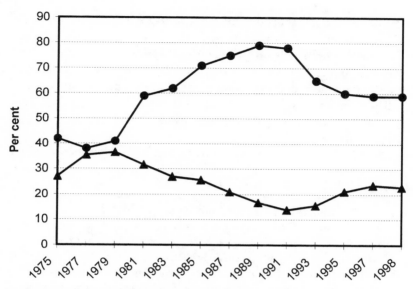

Figure 6.1. Changing patterns of consumption (triangles) and perception of marijuana as a health risk (circles) among 12th grade students in the United States (aged about 18). Consumption assessed as per cent that smoked during the past month. Data from United States National Institute on Drug Abuse (http://www.nida.gov).

the drug several times a day. If one looks at the NIDA data on American 18-year-olds in 1997, nearly 25% admitted to having used cannabis at least once during the past month and nearly 6% reported that they were daily users. The United States National Household Survey on Drug Use figures for 1992 suggest that 4% of the adult population (15–50 years old) were weekly users. In Canada 25% of children aged 15–17 were reported as current users (WHO, 1997). In Australia as many as 15% of men and 7% of women are weekly users (WHO, 1997). In most European countries around 5% of the adult population are current users (WHO, 1997). In Britain, data from the British Crime Survey 1994 for the adult population (aged 16–59), indicated that 5% had consumed cannabis during the past month. The WHO Report (1997) indicated that in Britain some 20% of young men (aged 16–24) and 12% of women in this age group were current users. It seems likely that only relatively small numbers of people on either side of the Atlantic use cannabis once a week or more, perhaps 4%–5% of the adult population, aged 15–50. Among young people regular consumption is much more common, with perhaps as many as 15–25% of 15–20-year-olds being regular users. The definition of "regular user," however, encompasses a wide range of consumption patterns.

Kate, a 25-year-old British cannabis user described her use of the drug as follows:

> I prefer cannabis to alcohol. It is so much more relaxing and social ; it is not like being in a pub with lots of loud music and drunk people being violent around you. The act of rolling a spliff puts an end to a working day, and marks the beginning of an enjoyable evening. The preparing of the joint, a complicated little process, carries with it a certain social ritual, which you do not get with alcohol — unless you mix a cocktail. I am never out of control. It is not a violent drug, all you do is get silly and start to recite old children's TV programs.

How Is Cannabis Consumed and Where Does It Come From?

In the United States the common form of the drug is herbal marijuana or sensemilla (dried female flowering heads), usually smoked with tobacco but also quite frequently on its own. The majority of users prepare their own hand-rolled joints, and more recently *blunts* — which are cigars emptied of their tobacco content and filled with marijuana.

The great majority of the supplies of this material during the 1970s were from cannabis grown on farms in the southern United States and in northern Mexico. The United States government's increasingly successful campaigns to eliminate these supplies (often by spraying the cannabis fields with the herbicide paraquat) lead to imports from further afield. Colombia and certain Caribbean countries, notably Jamaica, became more important. A report from the United States Department of State in 1997 estimated that there were 6500 hectares of cannabis still under cultivation in Mexico, 5000 hectares in Colombia, and 527 hectares in Jamaica, with the potential to produce a total of some 6000 metric tons of cannabis for export, worth several billion United States dollars (http://www.usis.usemb.se/drugs/Exec). In recent years there has been a large increase in the consumption of home-grown cannabis — often using modern strains of plants yielding a high THC content, and grown secretly indoors with artificial lighting. The Internet is a rich source of advice on how to grow cannabis at home, detailing the optimum heating and lighting needed and where to obtain the necessary seeds and equipment. An American television documentary recently described the new breed of marijuana farmers who grow the plant for sale. For an investment of less than $2000 they can obtain the equipment for a modestly sized indoor growing room, and with a crop cycle of only 6–12 weeks they can expect to make more than $100,000 a year of tax-free income.

> Americans are producing marijuana worth between $20 billion–$40 billion every year. We are talking about what may be the largest cash crop in America.
>
> (ABC News Saturday Night-"Pot of Gold"
> April 18, 1998)

The Independent Drug Monitoring Unit (IDMU) has provided some of the most detailed information available on the recreational use of cannabis in Britain. They undertook a number of surveys of cannabis users between 1982 and 1997, including those who attended outdoor pop music festivals. In their detailed evidence submitted to the House of Lords Cannabis Report (1998) the IDMU provided a summary of data received from 2794 regular drug users who completed their questionnaire. More than 90% of this group had used cannabis within the previous week, and more than 50% admitted to being daily users. In Britain as in the United States virtually all recreational use (96%) is by smoking, although in Britain the most common form of the drug is cannabis resin, which accounts for about 60% of consumption. The resin is most commonly smoked with tobacco in hand rolled joints or *spliffs* (70%). Some herbal cannabis is smoked without tobacco (5%), and resin is also smoked in pipes or bongs (16%). In addition, less common forms of consumption are also used by some. These include *hot knives,* in which a piece of cannabis resin is held between two heated blades and the resulting smoke inhaled; and the *bucket technique* in which the smoke from a smoldering piece of cannabis resin is captured in a bottle or bucket and then inhaled. A small proportion of users (4%) take the drug in food or drink, although as many as 25% of the IDMU group reported that they also did this occasionally.

Most cannabis resin in the United Kingdom is imported from Morocco or other parts of North Africa, with smaller amounts from Pakistan, Afghanistan, Lebanon, and the Netherlands. The cost of resin or herbal cannabis was about £4 ($6.70) per gram in most parts of Britain in 1997, and if the purchaser was willing to take the risk of being arrested in possession of a substantial quantity (for which the penalties are more severe), the price fell to around £2.50 ($4) per gram for a 9-bar block of resin weighing 250 g (about 9 ounces). The imported resin is often adulterated with other materials, commonly with caryophylline an aromatic constituent of cloves. The remainder of the cannabis consumed is herbal. About 10% of total consumption is in the form of compressed preparations of female flowering heads imported from Africa, the Caribbean, and the Far East. This material if often of poor quality, with molds growing on it. Because of the poor quality of imported supplies, homegrown cannabis is becoming more common and accounts for about 30% of total consumption. By growing such varieties as "Skunk" or "Northern Lights" under optimum growing conditions indoors it is possible to obtain uncontaminated herbal cannabis with a THC content of 10%–20%. The growers and their friends consume much of the homegrown cannabis, but some is commercially available. High potency herbal cannabis commands a premium price, up to twice that of the imported resin. British Home Office statistics report that police seizures of cannabis plants in Britain increased from 11,839 plants in 1992 to 116,119 in 1996.

Patterns of Recreational Use

Berke and Hernton (1977) undertook a questionnaire survey of 522 British cannabis users in the 1970s. Those surveyed were mainly young (16–25 years old) and predominantly students or recent graduates, although 191 (37%) were unemployed at the time of the survey. Most had used cannabis for at least 1 year, and about a third had been regular users for more than 5 years. The principal reasons they gave for starting to use cannabis were curiosity and social pressure. When asked why they continued to use cannabis the most common reasons given social uplift (pleasure, enjoyment, relaxation, increased sociability) (306 responses), a cheap and harmless alternative to alcohol and other drugs (167 responses), increased awareness and understanding (131 responses). And some said quite simply that they liked it (128 responses).

More recently the IDMU survey of regular cannabis users in Britain revealed a wide range of patterns of use (Atha and Blanchard, 1997; House of Lords, 1998). The average rate of consumption was 25–30 g per month (about 1 ounce). The average spliff contains 150–200 mg of cannabis resin, so average use equates to around 150–200 joints per month, or 5–7 each day. But the average figure conceals a wide range of levels of consumption; the median figure was 14 g per month. Among those admitting use more than once a day, the average consumption was 66 g of resin per month. The maximum levels of consumption reported were 150–200 g per month. The wide range of consumption levels is illustrated in Figure 6.2 and 6.3. It is clear that the great majority of users consume less than 15 g of resin per month and many do not take the drug on a daily basis. Herbal cannabis usually contains less THC, so consumption is higher, averaging 57 g per month. The wide range of cannabis consumption resembles that of alcohol, which is also consumed over a wide range of intakes, whereas the large majority of cigarette smokers fall within a narrow range of 15–40 cigarettes a day.

Neil Montgomery, a social anthropologist from Edinburgh, Scotland, gave evidence to the House of Lords (1998). He divided recreational cannabis users into three categories:

Distribution of monthly cannabis use

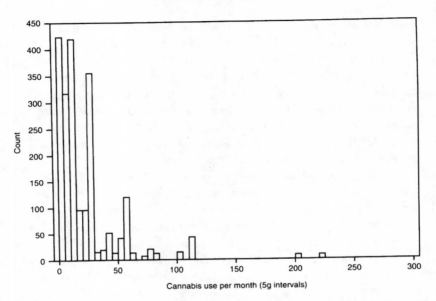

Figure 6.2. Distribution of monthly cannabis consumption in a group of 2469 regular cannabis users in Britain, surveyed between 1994 and 1997 by the Independent Drug Monitoring Unit. Data provided to the House of Lords Cannabis Report (1998). Reprinted with permission from I.D.M.U.

- Casual: Irregular use, in amount of up to 1 g resin at a time to an annual total of no more than 28 g (1 ounce).
- Regular: Regular use, typically 3–4 smokes of a joint or pipe a day, equivalent to about 14 g of cannabis resin (½ ounce) per month.
- Heavy: Only about 5% of total users, but they are more or less permanently stoned, using more than 3.5 g of resin per day and 28 g (1 ounce) or more each week.

His figures, based on his own research with more than 200 cannabis users, are consistent with those provided by the IDMU. Montgomery points out:

Distribution of number of 'spliffs' smoked per day

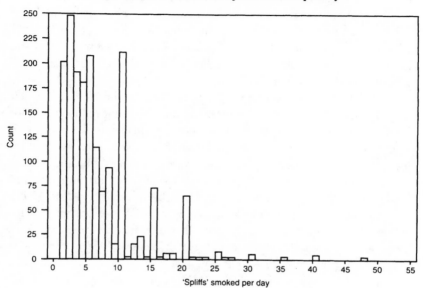

Figure 6.3. Distribution of number of marijuana cigarettes (spliffs) smoked per day in 2469 regular cannabis users in Britain, surveyed between 1994 and 1997. Data provided by the Independent Drug Monitoring Unit to the House of Lords Cannabis Report (1997). Reprinted with permission from I.D.M.U.

The extent to which a heavy user can consume cannabis is largely unappreciated. . . . These are people who have become dependent on cannabis; they are psychologically addicted to the almost constant consumption of cannabis. . . . Becoming stoned and remaining stoned throughout the day is their prime directive."

The maximum consumption figures reported by Montgomery and by the IDMU correspond to large intakes of THC. People consuming more than 5 g of cannabis resin a day may have a daily intake of as much as 200–300 mg of THC. These are by no means the highest figures reported in the literature, however. There are reliable records of people in the Caribbean consuming as much as 50 g of cannabis per day. One study of cannabis users in Greece estimated an average daily intake of 7.5 g (¼ ounce) a day. It is likely that such heavy users have become tolerant to most of the effects of THC. Montgomery estimated that the heavy users needed as much as eight times more cannabis than more modest consumers to become high.

The IDMU surveys refer mainly to young cannabis users, and illicit use continues to be far more common among those under the age of 30. Nearly all surveys also show that recreational cannabis use is about twice as common in men as in women. The IDMU noted, however, that although prevalence decreased in those over the age of 30, this may reflect more a cultural divide between generations and this situation may not hold in the future. They note that the British Crime Survey 1991–1996 reported the greatest proportional increases in cannabis exposure in older age groups. Lifetime prevalence more than doubled between 1991 and 1996 in those aged 40–44 (from 15% to 30%) and trebled in the 45–59 year-old-age group (from 3% to 10%).

Among those surveyed by the IDMU, as in other surveys of this type, the typical pattern of cannabis consumption for new users was an increase in the levels of consumption during the first 1–2 years (from an average of 12 g of resin per month in year 1 to 30 g in year 2) and a gradual decline to some stable level of consumption after year 5 (about 25 g per month).

What Are the Effects of Recreational Cannabis Use?

Harrison Pope and colleagues conducted anonymous questionnaire studies of illicit drug use at the same academic institution in the United States on three occasions over a 20-year period (1969, 1978, 1989), each time obtaining data from several hundred students (Pope et al., 1990).

Their results provide a valuable picture of marijuana use on the campus over this period. The incidence of marijuana use fluctuated widely, with weekly use reaching a peak of 26% of respondents in 1978 but falling to 5.7% in 1989. Marijuana was by far the most commonly used drug, followed by alcohol. There were no differences between drug-using versus nondrug-using students in most indices assessed; these included academic performance as measured by grade point averages, or participation in college athletic, social and political activities. Whereas drug users in the 1969 survey reported a significantly higher level of "alienation from American society," this was no longer true 20 years later. There were only two factors which distinguished drug users from non-users, they tended to visit a psychiatrist more often (although this did not seem to be directly attributable to drug use) and had more heterosexual experience (86% of the 1989 drug-using group reported having had intercourse with at least one partner, whereas only 52% of nonusers reported this).

Kandel et al. (1996) surveyed 7611 students, aged 13–18 in 53 New York schools. Of these, 995 had experience with marijuana, but there was no evidence that this had any significant impact on their school performance or their family relationships, whereas the small number (121) of crack cocaine users showed significant impairments in both.

In terms of adverse effects, Berke and Hernton (1977) in their survey of 522 British cannabis users found that about half of the group reported feeling physically ill on one or more occasions after taking cannabis, the most frequent symptoms being nausea, sickness, and vomiting. These symptoms occurred shortly after taking the drug and were transient (15–30 minutes). Dizziness, headache, or exhaustion were the next most frequent physical symptoms. A quarter of the group reported that on occasion they had unpleasant mental experiences. The most common symptoms were paranoia, fear, depression, anxiety, derealization, or hallucinations. A small number of people (49 of 522) admitted committing a socially irresponsible act while under the influence of the drug, the commonest being driving while stoned, fighting, or inappropriate sexual activity (although only 1% cited better sex as one of the effects of the drug).

In assessing what effect cannabis has on recreational users the IDMU survey data on British users are again a valuable source of information (Atha and Blanchard, 1997). While the majority of publications on this topic stress the adverse effects of cannabis, the overwhelming message from the British users was positive. When asked to rate their attitudes to a variety of psychoactive drugs, cannabis was given the highest positive rating, followed by ecstasy and d-LSD. Negative attitude ratings were given to solvents, cocaine, heroin and tranquilizers. Table 6.2 summarizes the positive benefits claimed by the regular cannabis users. The most common were relaxation, a sense of calm, and relief from stress. A variety of medical benefits were also reported, although only 2.8% of the group reported that medical use was their principal reason for taking the drug.

When asked why they took cannabis, more than 50% cited relaxation, pleasure, recreation, or social reasons. While the majority enjoyed the drug-induced experience, 21% of the group reported having experienced adverse effects on some occasion. These are summarized in Table 6.3. Adverse psychological effects were the most common. Very few users admitted to being dependent on cannabis.

Users were also asked whether they had ever been involved in road traffic accidents, and the results indicated that accident rates among this group of young people were not significantly different from the national rate for all drivers in this age group. The conclusion that cannabis does not appear to be a major cause of road accidents is, however, regarded as tentative pending further data.

Table 6.2. Most Common Positive Benefits Reported by 2794 British Cannabis Users

Effect	% Reporting
Relaxation/ relief from stress	25.6
Insight/personal development	8.7
Antidepressant/happy	4.9
Cognitive benefit	2.9
Creativity	2.3
Sociability	2.0
Health Effects	
Pain relief	6.1
Respiratory benefit	2.4
Improved sleep	1.6
Total reporting positive effects	57.8

Table 6.3. Adverse Effects Attributed to Cannabis by 2794 British Users

Effect	% Reporting
Impaired memory	6.1
Paranoia	5.6
Amotivation/laziness	4.8
Respiratory	4.2
Anxiety/panic	1.8
Cognitive impairment	1.7
Nausea	1.3
Dependence	0.6
Psychosis	0.4
Total Reporting Problems	21.0

The Potency of Illicit Marijuana

One of the claims frequently made by opponents of the recreational use of marijuana is that the cannabis used today is far more potent than the relatively harmless low-THC herbal material smoked by the flower power generation of the 1960s and 1970s. It is claimed that the supplies of cannabis available today are 10, 20, or even 40 times more potent than previously. As Professor Heather Ashton put it in her evidence to the House of Lords (1998) enquiry:

> The increase in potency is important because the physical and psychological effects of cannabinoids (THC and others) are dose-related: the bigger the dose the greater the effect. Most of the research on cannabis was carried out in the 1970s using relatively small doses, and much of that research is obsolete today. The acute and long term effects of the present high dose use of cannabis have not been systematically studied.

But is it really true that the commonly available cannabis today is more potent? And does it matter? For more than 20 years the United States government has sponsored the Potency Monitoring Project at the University of Mississippi that has been measuring the THC content of seized samples submitted by law enforcement agencies throughout the United States. At the meeting of the International Cannabinoid Research

Society in 1998, Dr. M. El Sohly from the University of Mississippi summarized the results obtained on more than 35,000 such samples since 1980. Marijuana leaf samples (the type most common in United States seizures) had around 2% THC content in 1980 and most recently the figures were 3.9% in 1996 and 4.1% in 1997. There was considerable fluctuation from year to year in the data — but it is clear that if there has been any progressive increase in the potency of herbal marijuana it represents not more than a doubling in THC content over nearly 20 years. The THC content in sensemilla (the female flowering heads) was around 6.5% in 1980, 9.2% in 1996, and 11.5% in 1997. Any increases in THC content were attributed to improved culture conditions rather than to any genetic improvements. Analysis of samples of cannabis resin or cannabis oil failed to show any discernible trends, with figures ranging from a 3% to a 19% THC content.

In Britain the United Kingdom government's Forensic Science Service provided data to the House of Lords (1998) enquiry on the THC content of cannabis samples seized in the United Kingdom (Fig. 6.4). They made the following statements:

> Cannabis resin, a wholly imported material, has a mean THC content of 4%–5%, although the range is from less than 1% to around 10%. This pattern has remained unchanged for many years.
>
> Herbal cannabis may be seen in a number of forms, but the material most commonly seized by Police and Customs in the UK has been imported in the form of compressed blocks; the mean THC content is also 4%–5% with a range similar to that of the resin.
>
> Until about eight years ago, "home grown" cannabis was a poor quality product often grown in greenhouses or on windowsills and normally for personal use. However, the introduction of a number of horticultural techniques has lead to the widespread and large scale domestic indoor cultivation of cannabis with a much higher THC content. These techniques include hydroponics, artificial lighting, control of "day" length, heating and ventilation, cloning of "mother plants", and perhaps most importantly, the development of plant varieties which produce higher THC levels. The mean THC content of so-called hydroponic cannabis is close to 10% with a range extending to over 20%.

The Tetrahydrocannabinol (THC) content of herbal cannabis (1996-98)

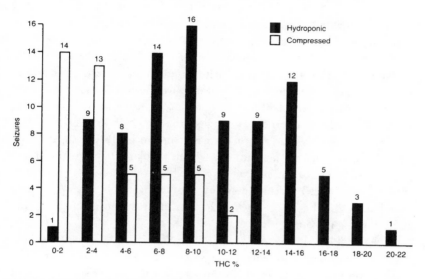

Figure 6.4. Tetrahydrocannabinol content of herbal cannabis samples seized by the police in Britain during the period 1996–1998. Data for imported (Compressed) herbal cannabis are shown separately from homegrown (Hydroponic) cannabis. Results provided by the United Kingdom Forensic Science Service to the House of Lords Cannabis Report (1998).

The conclusion on both sides of the Atlantic seems to be that the forms of cannabis that are commonly available commercially, herbal cannabis and cannabis resin, have changed relatively little in their potency over a period of some 20 years. The new strains of cannabis that have been bred for intensive indoor cultivation, with plants of short stature and high THC content, however, may be changing the picture. They yield herbal cannabis that contains 2–4 times more THC than has generally been available previously. It is also the case that such home grown material is becoming an increasingly important source of supply — accounting already for almost a third of United Kingdom consumption.

But is this necessarily a matter of concern? Looking at some of the positive aspects one could argue that if people are going to consume cannabis illegally, then is it not better that they consume material that has been grown under clean conditions? Such material is more likely than imported cannabis to be free of fungi or other microbial infections and is less likely to have been adulterated with other potentially toxic materials as commonly happens in imported cannabis resin. Because of the strictly controlled growing conditions hydroponic cannabis will tend to have a highly consistent THC content. Zimmer and Morgan (1997) have rehearsed the arguments for suggesting that high potency cannabis may not necessarily lead to an increased intake of THC. Experienced marijuana smokers are able to adjust their smoking behavior to obtain the desired level of high, and when offered high potency marijuana they inhale less smoke. From the point of view of the respiratory system one could argue that high potency THC is less likely to cause damage to the lungs for this reason.

Nevertheless, it is possible that the availability of these new forms of high potency marijuana will tempt some users to increase their THC intake, and this in turn could lead to a higher risk of dependency. With any psychoactive drug it is the users at the upper end of the consumption range who run the greatest risk of dependency. Reports that tobacco companies had developed genetically engineered strains of tobacco with an increased nicotine content, and that they had been using this material to bolster the nicotine content of some brands of cigarette were rightly met with consternation and suspicion (The Associated Press, High-Nicotine Leaf Used Despite Promises to Quit. *Newsday*, Feb 11, 1998).

In summary, the more extravagant claims about super-potent cannabis, suggesting that recreational users today are exposed to a wholly different drug from the one their parents may have consumed 20–30 years ago, are not supported by the evidence. On the other hand, hydroponic cannabis is a rapidly growing source of supply and it does contain a considerably higher THC content than has previously been available. Whether this is necessarily dangerous is not clear, it could increase the risk of dependency, but it may also be that the better consistency and quality of this product exposes users to less health hazards than before.

Is Marijuana a "Gateway" Drug?

A widely debated question is whether the use of marijuana leads people to use other illicit drugs, and eventually to become addicted to these drugs. Those who believe this to be true argue that even if marijuana is a relatively harmless drug it can act as a stepping stone to other far more dangerous drugs. In the 1960s the worry was that marijuana use might lead to LSD or heroin use. In the 1990s the principal concern is that it might lead to cocaine use.

> Although marijuana is not as addictive or toxic as cocaine, . . . smoking marijuana — or seeing others smoke marijuana — might make some individuals more disposed to use other drugs.
>
> (Chalsma and Boyum, 1994)

Many surveys have shown that young people who use psychoactive drugs begin with alcohol and tobacco and then marijuana. They tend to experiment with a number of other illicit drugs. Most who take cocaine will have had previous experience with marijuana and several other illicit substances. Kandel et al. (1996), for example, surveyed 7611 students aged 13–18 in 53 New York schools. Of the total, 995 had experience with marijuana, 403 had experience with cocaine, and 121 of these had taken crack cocaine. Alcohol or cigarette use tended to begin at age 12–13, marijuana use at age 15, and cocaine use at age 15–16. The young people who used drugs lived in social environments in which they perceived the use of drugs to be prevalent. Of the students who used crack cocaine, two thirds reported that all or most of their friends had used marijuana and 38% had used cocaine. Among nonusers of drugs, the corresponding figures were 8% and 0% respectively. But this does not prove that one drug leads to another, as Zimmer and Morgan (1997) point out:

> In the end, the gateway theory is not a theory at all. It is a description of the typical sequence in which multiple drug users initiate the use of high-prevalence and low-prevalence drugs. A similar statistical relationship exists between other kinds of common and uncommon related activities. For example, most people who ride a motorcycle (a fairly rare activity) have

ridden a bicycle (a fairly common activity). Indeed the prevalence of motorcycle riding among people who have never ridden a bicycle is probably extremely low. However, bicycle riding does not cause motorcycle riding, and increases in the former will not lead automatically to increases in the latter. Nor will increases in marijuana use automatically lead to increases in the use of cocaine or other drugs.

Kandel et al. (1996) found that parental behavior was an important determinant of the drug users behavior. Parental use of alcohol and cigarettes were important in determining experimentation with these drugs. Perhaps more surprisingly, parental use of a medically prescribed tranquilizer was likely to be associated with children's experimentation with illicit drugs. Through their use of legally available psychotropic drugs, parents may indicate to their children that drugs can be used to handle their own feelings of psychological distress.

So is the relationship that does exist between marijuana use and harder drugs simply a matter of social context? Is it the introduction to the underground world of illicit drugs through marijuana that leads people to experiment with other illicit substances? The Dutch believe that this relationship can be broken by separating the supply of hard drugs from that of marijuana, and making the latter freely available (Chapter 7). But is this really the whole story or might there be some neurobiological basis for the "gateway theory"?

Support for such a view seemed to come from basic research findings on the ability of THC to trigger activity in neural pathways in animal brains that use the chemical messenger dopamine (Fig. 3.2). The significance of these findings is that this is a common feature seen in response to a variety of addictive CNS drugs, including alcohol, nicotine, cocaine, amphetamines, and heroin. Some scientists have argued that it is the release of the chemical dopamine in certain key regions of the brain that is responsible for the rewarding effects of these drugs, and leads the user to wish to use them again. Others would argue that this is too simplistic, and that the significance of triggering dopamine release is that it may be "getting the brain's attention" to some significant stimulus (in this case the psychotropic drug), and that this in turn may be helping to determine the animal's motivation for seeking to repeat the experience. Furthermore, since alcohol and nicotine trigger dopamine release in the same way as THC, one could equally argue that these too should be considered "gateway" drugs to cocaine, heroin, or amphetamines. The results with THC were rendered more complex, however, by the finding that the effect of THC on dopamine release is apparently due to its ability to trigger a release of naturally occurring opioid substances in the brain (see Chapter 3). There seems also to be some crossover in the dependence syndromes caused by cannabinoids and opioids (Chapter 3). One could suggest that the reason some people become dependent on marijuana is because they can become addicted to their own naturally occurring opioid chemicals. Using marijuana may prime the brain to seek substances like heroin that act on the same opiate receptors.

These findings must give some pause for thought; they certainly cast a new light on the mechanisms involved in the actions of THC on the brain, and how they relate to opioid mechanisms. Brain researchers now see the endogenous opioid and cannabinoid systems in CNS as two independent but parallel and overlapping physiological regulatory systems. Both are involved in controlling our sensitivity to pain, and both may be involved in some way in the reward mechanisms in the brain. The subjective experience of taking marijuana is quite different from that induced by heroin or other opiate drugs. Experimental animals also find these drugs different; animals trained to discriminate THC or morphine do not mistake one drug for the other. There is also no evidence that administering THC makes animals more likely to self-administer heroin.

Nevertheless, the connection between opioid and cannabinoid mechanisms in the brain makes THC seem closer in some ways to morphine and heroin than any of the other psychotropic drugs that also trigger CNS dopamine mechanisms. This link to opioid mechanisms seems potentially far more significant than the link to dopamine mechanisms.

Do Recreational Marijuana Users Become Dependent?

As discussed in Chapter 3, there is evidence from animal studies that physical dependence to THC can occur, and there is human evidence of psychological dependence. A number of studies have shown that application of the internationally agreed DSM-IV (1994) criteria for substance dependence reveals that a substantial number of regular cannabis users can be classified as dependent, even though most would not admit to that description. Wayne Hall and Nadia Solowij, internationally recognized experts in the field of addiction research, describe how they view this situation:

> Dependence on cannabis is the most prevalent and under-appreciated risk of regular cannabis users. About 10% of those who ever use cannabis, and one third to one half of those who use it daily will lose control over their cannabis use and continue to use the drug in the face of problems they believe are caused or exacerbated by its use. . . . Uncertainty remains as to how difficult it is to overcome cannabis dependence and what is the best way to assist individuals to become abstinent.
>
> Hall and Solowij (1997)

Dr. Hall in testimony to the House of Lords Report (1998) said:

> By popular repute, cannabis is not a drug of dependence because it does not have a clearly defined withdrawal syndrome. There is, however, little doubt that some users who want to stop or cut down their cannabis use find it very difficult to do so, and continue to use cannabis despite the adverse effects that it has on their lives. . . . Epidemiological studies suggest that cannabis dependence, in the sense of impaired control over use, is the most common form of drug dependence after tobacco and alcohol, affecting as many as one in ten of those who ever use the drug.

Is this an exaggeration of the true situation, reflecting the particular problems that the authors see in their home country, Australia, which has a particularly high rate of cannabis consumption? One way of measuring the extent of cannabis dependence is by the number of people who seek treatment for it. In Britain, the 1996 Department of Health figures show that in 6% of all contacts with regional drug clinics, cannabis was the main drug of abuse. A similar figure, that cannabis users constitute 7% of all new admissions to drug treatment centers in Australia was also reported recently (House of Lords, 1998). In the United States NIDA reports that 100,000 people are enrolled for the treatment of cannabis dependency syndrome. It has been suggested that the United States figures are inflated by people on compulsory treatment after testing positive for cannabis at work, but the figures in Britain and Australia are unlikely to be influenced in this way as drug testing in the workplace is not yet common in either country. Chronic marijuana users who seek treatment report being unable to reduce their use of the drug, despite a strong desire to do so and despite the presence of distressing symptoms associated with continued use, such as diminished ability to concentrate, depression, and sleeplessness, suggesting addiction to the drug (Rohr et al., 1989; Tunving et al., 1995).

The Institute of Medicine Report (1999) suggested that 9% of those who ever used cannabis become dependent (as defined by the DSM-IV criteria); this compared with dependency risks of 32% for tobacco, 23% for heroin, 17% for cocaine, and 15% for alcohol.

Forensic Testing for Cannabis—Growth Industry of the 1990s

Cannabis is a potent drug so the concentrations of THC and THC metabolites in blood or other body fluids are very low. Whereas alcohol is present in quite high concentrations and is thus easy to measure in blood or breath, measuring cannabis proved technically much more difficult. Until the 1980s THC could only be measured in blood or urine samples after concentrating the sample and using complex chromatography equipment. The problem was solved by the development of immunoassay kits. These depend on using THC or its metabolite to stimulate the immune system of animals to produce antibodies that recognize THC or its major metabolite (carboxy-THC)(Fig. 2.7). Antibodies recognize the drug at very low concentrations, and they can be used as reagents in tests that involve measuring the binding of THC or carboxy-THC that triggers a change in fluorescence, a color reaction, or displaces a radioactive tracer. The availability of commercial kits for cannabis testing in urine has made such testing widespread. It is now routine to test road accident victims and hospital emergency room admissions for cannabis and a

range of other psychoactive drugs. The finding that a significant proportion of people involved in road traffic accidents or admitted to hospital
emergency rooms are testing positive for cannabis has been given much
publicity as another warning of the dangers of cannabis. But these com-
mentators ignore the fact that the use of cannabis is widespread, and
since the tests yield positive results for long periods after the last drug
use, it is not surprising to find many people registering positive. Drug
testing in the workplace has also become common — particularly in the
United States. Here the consequences of testing positive for cannabis
often can be severe — enroll in a cannabis treatment program and stop
using the drug or lose your job.

Because of the long persistence of THC in the body, cannabis tests
fail to give a reliable indication of the state of intoxication of the user. A
urine concentration of 50 ng/ml for carboxy-THC is generally taken as
the definition of a positive test. Such levels may occur in urine for days
or even weeks after the last dose of drug. By measuring the ratio of carboxy-THC to unchanged THC some idea can be obtained of how long
ago the last dose was taken — since this ratio increases with time. These
measurements, however, do not have the same value as measurements of
alcohol in breath or blood — which give a far more accurate picture of
the state of intoxication of the drinker at that moment. Forensic testing
with new techniques of gas chromatography linked to mass spectroscopy
gives an even more sensitive method for detecting minute quantities of
THC — levels of 1 ng/ml or less can readily be measured. These techniques can be applied to the analysis of the drug in hair samples — thus
indicating whether an individual is a chronic drug user, as the drug will
persist in tiny amounts in hair for long periods as it grows out.

Snapshots of Cannabis Use Around the World

Cannabis has been used for hundreds of years in different countries and
cultures both for recreational and medicinal uses, and also as an integral
part of religious rites (for reviews see Rubin, 1975; Robinson, 1996). An
understanding of this may help us to understand the modern vogue for

cannabis use in the Western world and to place this into a broader context. Many modern users of cannabis speak of their feelings of spirituality and oneness with God when intoxicated. Cannabis is used in many religions as a sacrament, such as in dagga cults in Africa, in Ethiopian Copts, as well as Hindus, Zoroastrians, Rastas, Buddhists, Taoists, and Sufis. Unlike drinking alcohol, the use of cannabis is not expressly forbidden in the Koran, and in many Moslem countries cannabis tends to take the place of alcohol.

India and Pakistan

The report of the Indian Hemp Commission (1895) gave a detailed account of the use of cannabis in the Indian subcontinent 100 years ago (see Chapter 7), and Chopra and Chopra (1957) described a situation that seemed to have changed little almost half a century later. There are two principal methods of consuming cannabis. Dried herbal cannabis, known as bhang may be chewed or eaten, or more commonly used to make a beverage often known as *thandai*. Many variants of this drink exist, bhang may be mixed with many other ingredients including milk, almonds, melon and poppy seeds, aniseed, cardamons, musk, and essence of rose. Sweetmeats containing bhang and even ice cream containing the powdered leaves may also be used. Whereas alcohol is generally looked down upon in Hindu society as it is considered taboo, the use of cannabis is socially sanctioned. Bhang is used in the Hindu religion in particular to celebrate the last day of the Durga Puja, and offerings of it are made to the god Shiva in Hindu temples. Bhang is also used by itinerant Hindu ascetics:

> Fulfilling a spiritual function . . ., the ascetics — called sadhus — radiate spiritual energy as they walk about the country, feeding the consciousness of India and the planet, and believe that the use of bhang supplies them with spiritual power, brings them closer to enlightenment, and honors Shiva, who is said to be perpetually intoxicated by cannabis.
>
> Voluntarily homeless, the sadhus live in the forest or in caves or walk perpetually, subsisting on alms. Their hair hangs in long matted strands, their skin is covered with dust or ashes, and they wear only a few rags or

nothing at all. Sadhus practice physical austerities including celibacy and long fasts without food or water. Bhang is said to help them center their thoughts on the divine and to endure hardships.

(Robinson, 1996)

Bhang may also be used on the occasion of other Hindu festivals, at marriage ceremonies, and other family festivities. The widespread use of bhang, however, has decreased markedly in India in this century. Bhang is probably equivalent to low grade marijuana in its THC content, and the watery infusions that are drunk probably contain rather little active drug, although milk (which contains fats) would be a more effective means of extracting THC. Intoxication after taking bhang is not common, and the Indian Hemp Drugs Commission's conclusion in 1897 that the moderate use of hemp drugs caused no appreciable physical, mental, or moral injury was probably correct. The other method of consumption involved smoking ganja (the compressed female flowering heads) or charas (cannabis resin), commonly in earthware pipe known as a chillum was regarded as potentially more harmful, and doubtless delivered more active drug. Smoking is always a communal activity, involving two to five people. Workmen, fishermen, farmers, and others who had to work long hours smoked cannabis to alleviate fatigue and relieve physical stress, often at the end of a working day. Sportsmen took it to improve their physical strength and endurance. Intoxication was rare, and most users were able to carry on their work or other activities. Ganja and charas smoking was generally looked down upon by the middle classes as a working class activity.

Nepal and Tibet

The advent of the hippie era and the migration of young Westerners to the Himalayas in search of cannabis and spiritual enlightenment lead to some remarkable changes in local attitudes to cannabis in these cultures. In Nepal, cannabis was traditionally used by Hindu yogis as an aid to meditation, and male devotees used it as a symbol of fellowship in their communal consumption of the drug. It was also used by older people to while away the time when they were too old to work in the fields. The advent of the hippie era and an influx of Westerners, however, brought about increased cultivation of cannabis, inflated prices, and a change in attitude of young, middle class Nepalese. Smoking cannabis came to be regarded as a novel, acceptable, and pleasurable mark of sophistication. This in turn lead a panic-stricken government in Nepal to introduce harsh new laws during the 1970s in an attempt to suppress the use of the drug (Fisher, see Rubin, 1975)

In Tibet, cannabis plays a significant role in some Buddhist ceremonies. According to Indian tradition and writings, Siddhartha used and ate nothing but hemp and its seeds for 6 years prior to announcing his truths and becoming the Buddha in the fifth century B.C.

Southeast Asia

Cannabis is common in Cambodia, Thailand, Laos, and Vietnam — many Americans were introduced for the first time to the drug during military service in Vietnam in the 1960s and 1970s. The plant tends to be cultivated on a family basis, with a few plants growing around the house. Herbal cannabis is freely available in markets, and is smoked together with tobacco. The herbal material is also used extensively in the local cuisine to impart an agreeable flavor and mild euphoriant quality to foods. Medically, cannabis is recognized as a pain reliever and is used in the treatment of cholera, malaria, dysentery, asthma, and convulsions. Cannabis is considered to be a source of social well-being, to be shared with friends and is also used to ease difficult work tasks (Martin, see Rubin, 1975)

Africa

The use of cannabis both for pleasure and for religious purposes is common throughout most of Africa, where it predates the arrival of Europeans. Known commonly as dagga, cannabis is a sacrament and a medicine to the Pygmies, Zulus, and Hottentots. Its use in religious ceremonies in Ethiopia is ancient, and it was taken up and used as a sacrament there by the early Coptic Christian church.

In Morocco, cannabis, known as kif, is traditionally served as a stimulant and as a means of relieving the pressures of daily life among the tribal groups living in the Rif mountains. The growing of cannabis in this northern region of the country has recently become an important agricultural export industry for an area that was previously the poorest agriculturally. In Morocco and other countries in North Africa many people maintain special rooms where kif is smoked while traditional stories, dances, and songs are passed to the young generation.

Caribbean and Latin America

Jamaica has become an important cultivation center for cannabis. The drug, known as ganja, was brought there by laborers from India in the midnineteenth century and spread to the black working class community where its use has become widespread. Ganja smoking is so prevalent among working-class males that the nonsmoker is regarded as a deviant. The occasion of first smoking attains the cultural significance of an initiation rite, and ideally should be accompanied by the ganja vision. Jamaica is also the home of a twentieth century religion known as Rastifarianism founded by Marcus Garvey in the 1930s in which cannabis plays a key role. Members of this religion known as Rastas accept some parts of the Bible, but believe that the Ethiopian Emperor Haile Selassie was a living God and represented "Jesus for the black race." Ethiopia is thought of as the ancient place of origin of black people, and an eventual return to Ethiopia would be their equivalent of nirvana. The ritual smoking of cannabis forms a key part of the Rastifarian religion; it is thought to cleanse both body and mind, preparing the user for prayer and meditation. Rastas, with their characteristic dreadlocks and their dedication to cannabis have permeated many aspects of modern culture, especially in the field of pop music. One of the most famous was the musician Bob Marley who died in 1981. In a song entitled "Kaya" written in 1978 he sang openly about marijuana (kaya is a Jamaican street term for marijuana):

> "Wake up and turn I loose
> Wake up and turn I loose
> For the rain is falling

> Got to have kaya now, got to have kaya now
> Got to have kaya now, for the rain is falling
>
> I feel so high, I even touch the sky
> Above the falling rain
> I feel so good in my neighbourhood
> So here I come again .....

The Zion Coptic Church is an American sect modelled on the Rastifarian movement in Jamaica; it too maintains a cannabis-based Eucharist. In 1989 Carl Olsen, a member of this church, sought to gain exemption from the cannabis prohibition laws in the United States by claiming the rights of church members under the First Amendment of the American Constitution to have the freedom to pursue their own religion, which in this case required the use of marijuana as a sacrament. The United States Drug Enforcement Agency won the case, pointing out among other things that Olsen's action in importing 20 tons of marijuana into the country seemed suspicious, as this was an outrageous quantity to supply the few hundred members of the church in the United States.

Cannabis smoking is common in many Latin American countries. Sometimes, as in Brazil, it was brought there by African slaves, and spread among the working people as "the opium of the poor." In Mexico and Colombia, the cultivation of cannabis for export has become an important cash crop, and along with this has come a widespread use of the drug. Whereas marijuana smoking in Colombia was formerly regarded as socially undesirable, it has become acceptable in many circles. Mexico has been the home of a number of religious sects that use cannabis as a sacrament. For example, a small community near the Gulf of Mexico use marijuana, which they call "la santa rosa," in their religious ceremonies. The dried herbal cannabis rests on the divine altar wrapped in small bundles of paper, along with artifacts of ancient local gods and images of Catholic saints. The men and women priests of the church chew small quantities of the herb and it gives them inspiration to preach to the congregation. The French anthropologist Louis Livet (see Walton, 1938) described a remarkable communal marijuana ritual among a sect of native Indians in Mexico. Participants were seated in a circle and each in turn took a puff at a large marijuana cigar, which he passed to his neighbor. The atmosphere at such meetings was joyful and filled with ritual chanting and convivial warmth. Each of those attending took a total of 13 puffs, and at the end consequently found himself in a state of hallucinatory excitement and intoxication. At the center of the circle was placed a sacred animal, an iguana. The animal attracted by the smell of the marijuana smoke also rotated 13 times, turning its head towards the cigar with its mouth open, inhaling the smoke. The animal was thought to represent the sacred incarnation of a god presiding over the ceremony, and when the iguana became intoxicated and fell down, the participants knew that it was time to stop passing the cigar! The reptile served a function akin to that of the pit canary in nineteenth century coalmines!

Conclusions

The recreational use of cannabis has become common in most Western countries. Up until now it has been an activity indulged in mainly by those under the age of 30, but this pattern may change as cannabis becomes more and more accepted as a part of our culture. It has been accepted and widely used, often as an alternative to alcohol, in many parts of the world.

There are health risks associated with cannabis use, particularly with smoked marijuana, but earlier reports of the dangers of cannabis have been proven to be exaggerated. There is a genuine risk of developing dependence on cannabis, and for some people it can come to dominate their lives and have a very negative impact. To many people it is regarded, rightly or wrongly, as a harmless weekend indulgence.

7

What Next?

I t is not the purpose of this book to persuade the reader to join one side or the other of the cannabis wars but rather to seek some middle ground in this debate that has become so polarized. By writing a book that attempts to take a cautious attitude toward the limited facts known about the subject, the author invites the criticism levied against Hall et al. (1994) for their balanced review *The Health and Psychological Consequences of Cannabis Use* prepared for the Australian Government. They were guilty of "harmful caution" (Ghodse, 1994). As Dr. Ghodse put it:

> The authors have been rigorous in making sure that their inferences are largely based on established evidence. This high degree of caution in interpreting evidence is commendable but has led to salient information on the probable health consequences of cannabis use that should be succinctly transmissible to the public, being diluted. This in turn has led to the presentation of an optimistic view of the consequences of cannabis use that renders the authors' apparent caution prejudicial, or even harmful.

By the end of twentieth century we have reached an interesting stage in the cannabis debate in the Western world. We must soon decide whether to reintroduce it into our medicine cabinets, and whether to accept, albeit grudgingly, that the recreational use of cannabis has become part of our culture. Can we learn something from the many reviews of these subjects that have been sponsored by governments and other bodies during the past century? Will scientific research on the newly discovered cannabinoid systems in the body help to produce new ways of using cannabinoids that avoid their intoxicating properties?

A Hundred Years of Cannabis Inquiries

A cynic might suggest that the decision to hold an expert inquiry into cannabis is the politician's way of avoiding having to debate the issue or to take any action. There have been expert inquiries around the world, nearly all have concluded that cannabis is a remarkably safe drug and many have recommended that limited medical use be permitted pending the outcome of the more detailed clinical research that is needed to approve properly sanctioned cannabis-based medicines. None of these inquiries led to any substantial legislative changes, but some of the more important ones are worth considering.

The Indian Hemp Drugs Commission Report (1894)

For a long time this was an obscure document, but it has rightly been given a new lease of life in recent years. This is a remarkable example of the manner in which the British Empire was governed in the nineteenth century. As Britain expanded her empire in India there was concern that the abuse of cannabis by the native peoples might be endangering their health. There were rumors that the asylums in India were filling with those driven insane by the abuse of cannabis. The late nineteenth century saw the British Parliament put in place new restrictions on the consumption of opium and alcohol in Britain. The Temperance League was formed to combat the evils of alcohol. One of the leaders of that movement raised a question in the British Parliament in March 1893 querying the morality of the trade in cannabis in India, a trade that was not only sanctioned by the British Indian administration but which also provided substantial tax revenues. The British Government requested the Government of India:

> . . . to appoint a Commission to inquire into the cultivation of the hemp plant in Bengal, the preparation of drugs from it, the trade in these drugs, the effect of their consumption upon the social and moral condition of the people, and the desirability of prohibiting the growth of the plant and the sale of ganja and allied drugs.

The commission, consisting of eminent British and Indian administrators and medical experts, reviewed the situation not just in Bengal but in all of British India. They undertook interviews with a total of 1193 witnesses in 13 different provinces or cities, using a standardized series of questions. The witnesses were carefully chosen to represent both the officials and a wide range of citizens. They were asked about the cultivation of hemp in their region, the preparation and consumption of hemp-related drugs, and the effects that the consumption of these were thought to have on the physical and moral well-being of the users. A particular question that the commission addressed was whether or not the consumption of cannabis led to insanity, as claimed by some. All of the mental asylums in British India were visited and the records of every patient claimed to be suffering from cannabis-induced psychosis carefully examined. The conclusion was that in most cases cannabis could not be held responsible, and in the few genuine cases of cannabis-induced psychosis the illness proved to be short-lived and reversible on stopping the use of the drug This conclusion is consistent with what most contemporary psychiatrists now believe. After 2 years of detailed and thorough work, the commission published its conclusions and the supporting data in a six-volume document in 1894. Its conclusions concerning herbal cannabis (bhang) (see Chapter 5) can be summarized as follows:

> . . . the Commission are prepared to state that the suppression of the use of bhang would be totally unjustifiable. It is established to their satisfaction that this use is very ancient, and that it has some religious sanction among a large body of Hindus; that it enters into their social customs; that it is almost without exception harmless in moderation, and perhaps in some cases beneficial; that the abuse of it is not so harmful as the abuse of alcohol; . . .

The commissioners were more circumspect about the smoked forms of cannabis, ganja and charas (Chapter 5). Several witnesses referred to the habit-forming properties of the smoked drug, a habit that was easy to form but hard to break. The commission, however, did not feel that prohibition was justified or necessary. Prohibition would in any case be difficult to enforce, would provoke an outcry from religious users, and might stimulate the use of other more dangerous narcotics.

In addition there was the question of what to do about alcohol:

> Apart from all this, there is another consideration which has been urged in some quarters with a manifestation of strong feeling, and to which the Commission are disposed to attach some importance, viz , that to repress the hemp drugs in India and to leave alcohol alone would be misunderstood by a large number of persons who believe, and apparently not

without reason, that more harm is done in this country by the latter than by the former.

The commission's report is remarkably sophisticated and surprisingly relevant to many of the issues debated in the present day "cannabis wars." The fact that several of the cannabis-producing regions of India received a significant part of their local government income from the revenues imposed by the British on the trade in cannabis products may have influenced the commission's benign conclusions, but it remains a thorough and objective analysis.

Mayor La Guardia's Report, *The Marihuana Problem in the City of New York* (1944)

Despite the passing of the Cannabis Tax Act in 1937, the illicit consumption of cannabis continued to grow in American cities, and it gained a notorious reputation in the media as a "killer drug"; a view encouraged by Harry Anslinger head of the Federal Bureau of Narcotics. In New York, Mayor La Guardia decided to try to find out just how harmful cannabis was. He appointed a committee of scientists to investigate; and the resulting investigation was the most thorough since the Indian Hemp Drugs Commission 50 years earlier. The committee organized clinical research on the effects of marijuana, using 77 prison volunteers (it was common practice at the time to use such volunteers as research subjects; most American pharmaceutical companies, for example, tested their new medicines on prisoners). The volunteers were given large doses of a cannabis extract or were allowed to smoke marijuana during a period of up to 1 month while in the Welfare Island Hospital. The doses of THC were unknown but must have been quite high since almost all of the subjects became high even at the lowest of the doses administered. The researchers were impressed with the low incidence of adverse side effects. The most common were anxiety (particularly among those subjects who had not used the drug before), nausea and vomiting, and ataxia (clumsiness). Nine subjects reported what were referred to as psychotic episodes, but these were all transient and were not considered serious. In addition, a careful comparison was made between a group of 60 prisoners on

Ward's Island who had been daily marijuana smokers and a group of nonusers. The investigators concluded:

> Prolonged use of the drug does not lead to physical, mental or moral degeneration, nor have we observed any permanent deleterious effects from its continued use.

As important as the clinical studies was the sociological research commissioned by the committee, using police officers in civilian clothes who lived in the areas of the city in which marijuana use and peddling were common. The question of how widespread marijuana use was among school children was also addressed. The investigators concluded that marijuana use was largely confined to the poorer communities in the city, particularly in Harlem, that there was no link between marijuana use and crime and that the drug did not provoke violent behavior. There was no evidence of widespread use among school children. Furthermore, the report concluded:

> We have been unable to confirm the opinion expressed by some investigators that marijuana smoking is the first step in the use of such drugs as cocaine, morphine and heroin. The instances are extremely rare where the habit of marijuana smoking is associated with addiction to these other narcotics.

But although Mayor La Guardia's (1944) report was one of the clearest and most thorough investigations ever undertaken, its conclusions did not make much impression on public opinion in America at the time. The conclusion that marijuana was a relatively harmless drug was not what the media or Harry Anslinger wanted to hear. Anslinger was harshly critical of the report's conclusions, and even the influential *Journal of the American Medical Association* attacked the report in an editorial, which concluded:

> Public officials will do well to disregard this unscientific, uncritical study and continue to regard marijuana as a menace wherever it is purveyed.

Jerome Himmelstein (1978) in his book *The Strange Career of Marihuana* gives some remarkable insights into the strange history of the politics and ideology of cannabis in the United States. Detailed accounts of this history can also be found in Abel (1943) Robinson (1996), and Bonnie and Whitbread (1974). Public perceptions of the drug owed little to a dispassionate review of the scientific facts, and much more to the dedicated anticannabis crusade of Harry Anslinger and his Federal Bureau of Narcotics, and the popular disapproval of marijuana as a drug associated with the lower classes and with Mexican immigrants.

The Wootton Report, England 1968

The widespread consumption of cannabis did not begin in England or in most West European countries until the 1960s. Attitudes toward the control of the drug until then were driven largely by events across the Atlantic and by the various international agreements that were put into place, starting with the League of Nations Opium Conference in 1925, which categorized cannabis along with opium as a dangerous narcotic, and the World Health Organization Single Convention on Narcotic Drugs adopted in 1964, which similarly categorized cannabis as a Schedule I drug of addiction with no medical uses.

It was only when the use of cannabis suddenly expanded in the 1960s that the government at the time felt any need to take it more seriously. The British Home Office, in charge of the regulation of illicit drugs, established a group of experts known as the Advisory Committee on Drug Dependence, and an expert subcommittee of this was set up "to review available evidence on the pharmacological, clinical, pathological, social and legal aspects of these drugs (cannabis and lysergic acid)." An experienced sociologist and politician, Baroness Wootton, chaired the subcommittee. While the subcommittee was deliberating, an advertisement appeared in the *London Times* on July 24, 1967 asserting that the dangers of cannabis use had been exaggerated and advocating a relaxation of the laws governing its consumption. This provoked a wave of debate in the media and in Parliament. The *Wooton Report*, as the document submitted to the Home Secretary, James Callaghan in 1968 became known, made a big impact (Advisory Committee on Drug Dependence, 1969). Its conclusions were clear:

We think that the adverse effects which the consumption of cannabis in even small amounts may produce in some people should not be dismissed as insignificant. We have no doubt that the wider use of cannabis should not be encouraged. On the other hand, we think that the dangers of its use as commonly accepted in the past and the risk of progressing to opiates have been overstated, and that the existing criminal sanctions intended to curb its use are unjustifiably severe.

The report went on to recommend a number of changes to the criminal law, the chief of which would have made the possession of small amounts of cannabis for personal use no longer an imprisonable offense, but merely punishable by a summary fine. In addition it recommended that preparations of cannabis should continue to be available for medical uses. But like the La Guardia report earlier, *The Wootton Report* was assailed in the press and parliament as a "charter for drug seekers." By the late 1960s the large-scale spread of cannabis use on both sides of the Atlantic to middle class youth altered public perceptions of the problem. Cannabis use had become a symbol in the public mind of the hippie counterculture and the increasing alienation of young people from society. Perhaps these considerations lead the British Home Secretary James Callaghan to dismiss the *Wootton Report* in a statement to Parliament shortly after the publication of the report:

I think it came as a surprise, if not a shock, to most people, when that notorious advertisement appeared in the Times in 1967, to find that there is a lobby in favour of legalising cannabis . . . it is another aspect of the so-called permissive society, and I am glad if my decision has enabled the House to call a halt to the advancing tide of permissiveness.

Report Followed Report

At about the same time *The Wootton Report* was published in England, the United States Department of Health, Education and Welfare launched an ongoing study of implications of marijuana use in the United States through a National Commission on Marihuana and Drug Abuse. The first of a series of reports entitled *Marihuana: A Signal of Misunderstanding* (National Commission, 1972), (sometimes referred to as the *Shafer Commission Report*), produced a great impact. It went even further than *The Wootton Report* in recommending that the private possession or distribution of small quantities of cannabis for personal use should no longer be an offense, and that possession in public of up to 1 ounce (28 g) be punishable by a fine of $100.

> . . . marihuana use is not such a grave problem that individuals who smoke marihuana, and possess it for that purpose, should be subject to criminal prosecution.

Predictably, President Nixon summarily rejected these recommendations and there was a hostile reaction from many other quarters. One year later the commission published a second report *Drug Use in American: Problem in Perspective* (National Commission, 1973), which backtracked on the earlier recommendations:

> The risk potential of marihuana is quite low compared to the potent psychoactive substances, and even its widespread consumption does not involve the social cost now associated with most of the stimulants and depressants. . . . Nonetheless, the Commission remains persuaded that availability of this drug should not be institutionalized at this time . . . it is painfully clear from the debate over our recommendations that the absence of a criminal penalty is presently equated in too many minds with approval, regardless of a continued prohibition on availability. The Commission regrets that marihuana's symbolism remains so powerful, obstructing the emergence of a rational policy.

In Canada the *La Dain Report* (Canadian Government, 1970) provided a detailed review of cannabis use and it too recommended a repeal of the prohibition against the simple possession of cannabis. The Canadian authors also concluded that there was little evidence that cannabis was a drug of addiction. Like other reports published at that time the Canadian Commission found little to worry about:

> On the whole, the physical and mental effects of cannabis, at the levels of use presently attained in North America, would appear to be much less serious than those which may result from excessive use of alcohol.

In Australia and New Zealand, a report on drug trafficking and drug abuse published in 1971 revealed that cannabis use was increasing rapidly in that part of the world, with its favorable climate for cannabis cultivation. The authors did not appear to be alarmed by this, and recommended that first-time offenders no longer be subject to prison sentences but be given suspended fines.

The early 1970s represented the zenith of acceptance of marijuana as a relatively safe drug. The various groups of experts around the world who reviewed the subject helped to demolish the commonly held view that cannabis was a highly dangerous drug that rapidly produced disastrous effects on the mental and physical health of users. The example of tobacco smoking, where 20 years or more of continuous exposure are needed before the serious health consequences are seen was almost forgotten in the wave of euphoria for cannabis. For a while in the 1970s it looked as if the decriminalization of cannabis might be approved in the United States and elsewhere around the world. President Jimmy Carter was reported to be in favor of decriminalization and to have said that:

> Penalties against a drug should not be more dangerous to an individual than the use of the drug itself, and where they are they should be changed.
> (Zimmer and Morgan, 1997)

During the late 1970s and 1980s, however, an active antimarijuana movement gained ground, particularly in the United States. The arguments against the drug were largely moral, and were lead by politicians and by those scientists and psychiatrists who were willing to disclose only the adverse effects of the drug. Professor Gabrial Nahas, a scientist now at New York University was a particularly vocal and unashamedly biased campaigner against cannabis. His books *Marihuana — Deceptive Weed* (1973) and *Keep Off the Grass* (1976) helped to inflame if not to illuminate the debate. This campaign was joined also by well-meaning and well-organized groups of middle class parents who had no direct experience of cannabis but feared the dangers it might hold for their children. The National Institute on Drug Abuse also became more and more actively involved in publicizing the dangers of cannabis use, and continues to do so today (Zimmer and Morgan, 1997).

In the United States and most other countries in the Western world an impasse was reached. Criminal sanctions prohibiting the use of cannabis remained in place, although this seemed to have relatively little effect on the consumption of the drug, which continued to involve large numbers of young people. Cannabis was also finally excluded altogether from any medical uses, although as described earlier it was the revival of interest in this aspect of the drug that has rekindled the cannabis debate in recent years.

There have been several more recent reviews of the physical and mental consequences of cannabis use, including the excellent and thorough review by Wayne Hall and colleagues for the Australian Government, *The Health and Psychological Consequences of Cannabis Use* (Hall et al., 1994). Most recent was the WHO report, *Cannabis: A Health Perspective and Research Agenda* (WHO, 1997). But these reviews largely depended on research done in the 1960s and 1970s — the field of cannabis research was relatively dormant during the 1980's. It has come alive again in the 1990's with the new scientific discoveries of cannabinoid receptors and endogenous cannabinoids, and the increasing interest in the medical applications of cannabis.

The Dutch Experiment

Only one country in the West, Holland, decided to decriminalize cannabis. For the past 20 years the Dutch have taken a radically different approach in their drug policy (for review see Engelsman, 1989). The Netherlands signed the United Nations Single Convention on Narcotic Drugs (1964) and Dutch law states unequivocally that cannabis is illegal. Yet in 1976 the Dutch adopted a formal policy of nonenforcement for violations involving possession or sale of small quantities of cannabis (originally 30 g, reduced to 5 g since 1995). A group of Coffee Shops was licensed to sell small quantities of herbal cannabis or cannabis resin for consumption on the premises or to take away. The number of such establishments was small, however, until the late 1980s and 1990s. Now more than a thousand such establishments exist in the Netherlands. They must not hold more than 500 g cannabis in stock, are not permitted to sell alcohol or any other psychoactive drugs, must not cause a nuisance to neighbors, cannot advertize, and are not permitted to sell cannabis to minors. These regulations are strictly policed and licenses can be revoked and the owners punished for violating them. The aims of Dutch drug policy are pragmatic rather than moralistic, they hope to achieve "harm reduction" by regulating the traffic in cannabis and separating this from the sources of supplies of other illegal and potentially more harmful psychoactive drugs. A saying can sum up this attitude:

We don't solve a problem by making it taboo and pushing it underground.

But have the objectives of Dutch cannabis policy been achieved? Many critics from outside the country portray lurid tales of decadence and cannabis-doped youth. What are the facts? Did the levels of cannabis use increase rapidly after decriminalization in 1976? Are the levels of cannabis use higher in the Netherlands than in other Western countries? The best available comparisons of data on cannabis consumption among 18–20 year olds show that the new policy had surprisingly little impact on cannabis consumption among young people in Holland, which remained stable for some years after the new policy was introduced until it started to rise in the mid 1980s (MacCoun and Reuter, 1997). Between 1984 and 1996 the use of cannabis in Holland increased rapidly, with lifetime exposure in the 18–20-year-old group rising from 15% in 1984 to 44% in 1996, and exposure during the previous month rising from 8.5% to 18.5%. However, similar rapid increases in cannabis consumption in this age group were observed during the 1990s in the United States and in Norway, two countries that have strictly enforced prohibition laws. There is some evidence that the Dutch consumption of cannabis rose faster during the 1980s than elsewhere, probably as a result of the "coffee shop policy." The fact remains that the current levels of cannabis use among young people in Holland are comparable to those in other European countries and lower than those in the United States, even after 20 years of decriminalization. Whether the Dutch experiment has succeeded in its objective of separating the use of soft and hard drugs is less easy to answer. There are some positive data, for example, the average

age of heroin addicts in Holland is increasing—suggesting that fewer young people are being recruited to heroin addiction. In 1981, 14% of Dutch heroin addicts were under 22, today the figure is less than 5%. In 1995, the number of heroin addicts per 100,000 population was 160 in Holland versus 430 per 100,000 in the United States. But there remains an association between cannabis use and exposure to other psychoactive drugs—cannabis users are far more likely to have experimented with other psychoactive drugs than nondrug users. It is perhaps too early to say how successful the experiment has been in this regard (Ossebaard, 1996).

The Dutch approach would not fit easily in many other countries. It requires an ability to look the other way, which others might find more difficult. The coffee shop customers come through the front door and purchase small amounts of cannabis with impunity, but the coffee shop owner has no legal source of supply. He must obtain supplies of cannabis where he can, and have them delivered through the back door. More than half of all cannabis consumed in Holland is home-grown—with increasing horticultural expertise and new strains of high THC-content cannabis plants. The rest is imported, mainly from Morocco. But the suppliers are still liable to severe penalties if caught. Other European countries have complained that Holland has become an easy source of supply of cannabis for drug tourists from all over Europe who may carry away their purchases with little risk across a European Union that no longer has many border controls. Public opinion in Holland is by no means unanimously in favor of the present relaxed drug laws. Some aspects of Dutch policy are also hard to understand: while medical cannabis enjoyed a boom in the past decade in Holland, legal sources of supply for medical cannabis were not approved by the government, who did not admit that the drug has any legitimate medical uses. Paradoxically, the health council of the Netherlands issued a report in 1996, which concluded that there was insufficient evidence to justify the medical use of marijuana. This situation may now be changing as a move to form a government agency to supply medical marijuana was recently announced.

The Dutch experiment has not been repeated anywhere else so far, although some States in the United States decriminalized cannabis possession for a while in the 1970s. As in the Netherlands this did not seem to lead to any marked increase in cannabis consumption (Institute of Medicine Report, 1999). The possession of small amounts of cannabis for personal use is also no longer punished in Spain or Italy, or in some regions of Australia. The country most likely to follow the Dutch in permitting the sale of cannabis from licensed premises is Switzerland, where moves are being made to adopt a policy of separating the supply of hard and soft drugs. France, Germany, and Britain remain firmly attached to their present policies of prohibition and punishment, as does the United States.

The Campaign for Medical Marijuana

As described in earlier chapters, many lines of evidence suggest that cannabis and cannabinoids have a number of valid medical applications. The immediate problem facing anyone wishing to use, or even to do research, on cannabis in medicine, however, is that it is categorized as a Schedule I narcotic, i.e., a dangerous drug with no medical utility. This means that in order to use it at all, even for research purposes, permission needs to be obtained from the appropriate department of government responsible for drug control. In the United States this is the Drugs Enforcement Agency (DEA), in Britain the Home Office. The United States government supported research on the medical uses of cannabis during the 1970s and made standardized marijuana cigarettes available for research purposes. Some patients were even able to obtain compassionate supplies of marijuana from the government. But this all came to an end under the Reagan and Bush administrations in the 1980s when attitudes to marijuana hardened. In 1972, in a famous test case, the National Organization for the Reform of Marijuana Laws (NORML) petitioned the DEA to reclassify marijuana as a Schedule II drug. That would permit physicians to prescribe it to their patients for compassionate use on a case by case basis. It was only in 1986 after years of legal wrangling that the DEA acceded to the demand for public hearings on this petition. The hearings lasted for more than 2 years and thousands of pages of evidence were accumulated. But the DEA denied the petition, even though their own legal expert Judge Francis L. Young recommended that the rescheduling be allowed. Judge Young concluded that marijuana had been shown to be "one of the safest therapeutically active substances known to man."

Nevertheless, there was a growing illicit medical use of marijuana — especially by patients suffering from AIDS — and doctors were inevitably aware of this and often connived in the use of the drug by their patients. In 1995, the *Journal of the American Medical Association* published an invited editorial by the long-standing advocates of medical marijuana, Lester Grinspoon and James Bakalar. They wrote:

> The American Medical Association was one of the few organizations that raised a voice in opposition to the Marijuana Tax Act of 1937, yet today most physicians seem to take little active interest in the subject, and their silence is often cited by those who are determined that marihuana shall remain a forbidden medicine. Meanwhile, many physicians pretend to ignore the fact that their patients with cancer, AIDS or multiple sclerosis are smoking marijuana for relief, some quietly encourage them. In a 1990 survey, 44% of oncologists said they had suggested that a patient smoke marihuana for relief of the nausea induced by chemotherapy. If marihuana were actually unsafe for use even under medical supervision, as its Schedule I status explicitly affirms, this recommendation would be unthinkable. It is time for physicians to acknowledge more openly that the present classification is scientifically, legally and morally wrong."

In November 1995, the authoritative British medical journal *The Lancet* (November 11, 1995, p.1241) also published an uncharacteristically hard hitting editorial entitled "Deglamorising Cannabis" which began:

> The smoking of cannabis, even long term, is not harmful to health."

The editorial went on to call for the decriminalization of cannabis along the lines of the Dutch experiment.

In November 1996, voters in California and Arizona approved propositions to make marijuana legally available for medical use. The state government in Arizona made it clear that it had no intention of acting upon the voters' wishes, but this was not the case in California. This led to an ongoing dispute between the state government and the federal government who maintained that cannabis was an illegal drug and could not be provided for any purpose. Doctors were threatened with having their medical licences revoked if they cooperated with the California scheme. Nevertheless cannabis buyers clubs were established and for a while they supplied patients with cannabis for medical use. Some cannabis clubs, however, were not as cautious and seemed intent on making the drug available to all comers. Most notoriously in San Francisco, marijuana guru Dennis Peron served approximately 10,000 clients from the San Francisco Cannabis Cultivators Club near the city's civic center. The club was decorated with thousands of origami birds dangling from mobiles, brilliantly colored jungle murals, and was permeated with hard rock music and the unmistakable smell of marijuana smoke. Peron was quoted as saying:

"All marijuana use is medicinal."

The Institute of Medicine Report (1999) included survey data on members of the San Francisco Club and the similar Los Angeles Cannabis Resource Center. Almost two-thirds of the patients surveyed at each location were using marijuana to treat HIV or AIDS-related symptoms and the remainder suffered from a wide range of conditions, including cancer, musculoskeletal disorders (arthritis, rheumatism, multiple sclerosis), chronic pain, and mood disorders.

After a considerable legal battle between the state and the federal governments, the buyers clubs were forced to close, leaving medical users with no organized means of supply. In January 1997, the prestigious medical publication *The New England Journal of Medicine* published an unusual editorial entitled *Federal Foolishness and Marijuana* which criticized the United States government's attempts to quash the California initiative and argued for the rescheduling of marijuana.

Meanwhile a number of expert groups were set up to review the case for the medical use of cannabis, and their findings have already been referred to elsewhere in this volume. In the United States, the American Medical Association report *Medical Marijuana* (1997) recommended that controlled clinical trials be undertaken with smoked marijuana in a variety of conditions, but that the drug should remain in Schedule I pending the outcome of such studies. The United States NIH *Report on the Medical Uses of Marijuana* (1997) also advocated clinical trials with smoked marijuana, and pointed out that such research was hindered by the Schedule I status of the drug. In Britain, the British Medical Association (BMA) published a review of the *Therapeutic Uses of Cannabis* (1997) which advocated further clinical research on the medical uses of synthetic cannabinoids. The BMA report was opposed to any use of herbal cannabis, however, on the grounds that it contained too many potentially active ingredients and could not be standardized. The House of Lords Science and Technology Committee report *Cannabis. The Scientific and Medical Evidence* (1998) also recommended that more clinical research was needed on both synthetic cannabinoids and herbal cannabis, and argued in favor of rescheduling cannabis to Schedule II to permit the compassionate use of the drug by doctors on a named-patient basis. The British government speedily rejected this recommendation.

The United States National Academy of Sciences, Institute of Medicine published the most comprehensive and complete review in the spring of 1999. This review took 2 years to complete and involved a series of discussion meetings around the United States. The Institute of Medicine report was particularly important as it was commissioned at the request of the Untied States drug czar, General McCaffrey. It concluded that there were genuine scientific grounds for exploring the therapeutic applications of cannabis, and listed pain relief, control of nausea and vomiting, and appetite stimulation as the top priorities. The report called for research on improved delivery systems, and like the House of Lords (1998) in the previous year, recommended the approval of smoked marijuana on compassionate grounds for certain categories of patients.

Those who oppose any change in the status quo of the regulations governing cannabis use view the moves to approve cannabis-based medicines as dangerous. Some see the campaign for medical marijuana as part of an overall campaign by some groups to legalize cannabis. As drug czar Barry McCaffrey put it in a press release, November 15, 1996:

There could be no worse message to young people. . . . Just when the nation is trying its hardest to educate teenagers not to use psychoactive drugs, now they are being told that marijuana is a medicine.

But these arguments ignore the fact that such powerful opiate drugs as heroin and morphine have a genuine and important place in medicine, despite their known dangers of abuse. There is no evidence that the medical use of opiates creates addicts, or that there is any substantial diversion of medical supplies of opiates to the illicit market. Given strict controls and the proper education of doctors in how to use the drugs they pose no particular health or psychiatric risks. The same could be true for cannabis and cannabinoids in their controlled use in medicine. As the Institute of Medicine (1999) report put it:

> . . . there is a broad social concern that sanctioning the medical use of marijuana might increase its use among the general population. At this point there are no convincing data to support this concern. The existing data are consistent with the idea that this would not be a problem if the medical use of marijuana were as closely regulated as other medications with abuse potential.

The debate on medical marijuana has reached an interesting and critical stage. In the United States voters in the November 1998 elections in Arizona, Alaska, Washington, Oregon, Nevada, Colorado, and the District of Columbia approved medical marijuana. Alaska and Washington were the first to put such a law into effect early in 1999, despite hostility from the federal government. The medical marijuana laws in these states would permit patients who have their doctors' approval to possess small amounts of cannabis or up to six or seven cannabis plants without penalty. No doubt the federal government will contest the implementation of these laws, as it did in California. More than 20 state legislatures have enacted some form of medical marijuana legislation, although it is not clear how these proposals can be put into effect in the face of implacable opposition from the federal government.

Perhaps most hopeful are the initiatives that are being taken on both sides of the Atlantic to mount large scale controlled clinical trials, using both synthetic cannabinoids and herbal cannabis. Some of these trials in the United States will involve smoked marijuana. In Britain a working party established by the Royal Pharmaceutical Society in 1998 published its recommendation on the design of two large clinical trials early in 1999. One would be in patients with postoperative pain, the other in patients with multiple sclerosis. Provided adequate sponsorship can be obtained, each trial will involve several hundred patients and will compare placebo with synthetic THC (dronabinol) or a herbal cannabis extract containing a comparable amount of THC in a double-blind randomized manner. Such trials are the only way in which official regulatory approval will be granted eventually to cannabis-based medicines. The present plans still face a number of obstacles. Because of the Schedule I status of cannabis, legal permission has to be obtained as well as the usual clinical trials approval procedures. It will take at least 5 years before any cannabis based medicine could be formally approved as a result of these moves — but at least they represent a concrete start.

What Promise Does Cannabis Research Hold?

The recent upsurge of scientific interest in cannabis and the physiological cannabinoid control system in the body looks as though it will continue. Scientists are excited by the discovery of this new natural regulatory mechanism and are eager to explore it. From this research there could be many spinoffs. The discovery of a second cannabinoid receptor, CB-2, present outside the brain in the immune system already showed how fundamental discoveries may point to completely new medical uses of cannabis-based drugs — which lack any psychoactive effects. It is possible that the two cannabinoid receptors and the two endogenous cannabinoid molecules that we know of so far represent only the tip of an iceberg, and that many other endogenous cannabinoids and receptors remain to be discovered.

We have to admit though that most and perhaps all of the therapeutic indications for which cannabis is presently being pursued seem to be based on the actions of THC or related cannabinoids on the CB1 receptors in the CNS. The availability of SR141716A and other drugs that act as antagonists at the CB1 receptors has helped to make this clear. No way has yet been found to obtain the beneficial therapeutic effects of cannabinoids without approaching doses that lead to their unwanted psychoactive effects. Indeed it is hard to see how cannabinoid drugs that act directly to stimulate the CB1 receptor will ever achieve this separation. But there may be other ways of manipulating the function of the cannabinoid system in the CNS. It is possible, for example, that the activity of anandamide and other endogenous cannabinoids in the brain could be enhanced by using drugs that blocked their normal inactivation by enzymatic breakdown or by tissue uptake mechanisms. This would allow a more subtle up-regulation of cannabinoid activity in the CNS since the effect would be greatest in those regions of the brain in which there was the greatest ongoing release of endogenous cannabinoids. By analogy with existing psychoactive drugs, one could compare the serotonin-uptake inhibitors (of which Prozac® is the best known example) with drugs such as d-LSD that acts directly on one of the serotonin receptors in brain. While Prozac® enhances serotonin function and is a valuable antidepressant, d-LSD is a powerful hallucinogenic agent of no medical value. It is possible that inhibitors of anandamide uptake could prove far more useful than THC or other drugs that act directly on cannabinoid receptors.

Another way in which research could assist progress in the medical applications of cannabis-based drugs would be in devising novel methods for delivering THC, synthetic cannabinoids, or herbal cannabis more efficiently. The limitations of the oral routes have frequently been referred to, but smoking despite its effectiveness as a means of drug delivery carries too serious a risk of respiratory disease and cancer to be a viable means of delivery, except for very short-term uses. Some research has already been done on a variety of vaporizer or other inhalation devices that would avoid the toxic ingredients in marijuana smoke, but these are still far from satisfactory. The delivery of THC by inhalation as an aerosol to the lung has received little attention, and the application of modern inhalation device technology could prove rewarding. Delivery of cannabinoid drugs to the lungs could potentially be as rapid and effective as smoking.

Large-scale epidemiological research on the long-term risks associated with marijuana smoking is also badly needed. The wave of cannabis research during the 1970s was unusual in that much of it was done with human subjects. There has been much less of this kind of research in recent years, but the questions have not all been answered. The various reports issued around the world during the 1970s on the health hazards of cannabis focused largely on the immediate effects of the drug. There have been few studies in Western societies that have followed up people who have smoked marijuana for 20 or 30 years. By analogy with tobacco smoking, it is only by doing this that an answer can be obtained on the long-term risks of marijuana smoking — particularly in terms of its potential to cause lung cancer.

Historical Changes in Attitudes to Psychoactive Drugs

In the often emotional debate regarding cannabis it may help to remember what remarkable changes there have been in attitudes to psychoactive drugs in Western society. A substance that is considered safe and beneficial in one era may be seen as an evil scourge to another generation.

Opium

A hundred years ago the dangers of opiate drugs were only just becoming recognized. Morphine and heroin have become widely abused during the twentieth century, and continue to wreak serious damage on the fabric of society in many Western cities. But opium, the crude resin of the opium poppy, which is rich in morphine, was freely available and widely used in English society in the nineteenth century, as described by Virginia Berridge in her fascinating book *Opium and the People* (Berridge and Edwards, 1981). They describe the variety of opium products that were freely available in England by the middle of the nineteenth century:

> The opium preparations on sale and stocked by chemist's shops were numerous. There were opium pills (or soap and opium), and lead and opium pills, opiate lozenges, compound powder of opium, opiate confection,

opiate plaster, opium enema, opium linament, vinegar of opium and wine of opium. There was a famous tincture of opium (opium dissolved in alcohol), known as laudanum, which had widespread popular sale, and the camphorated tincture, or paregoric. The dried capsules of the poppy were used, as were poppy fomentation, syrup of white poppy and extract of poppy. There were nationally famous and long established preparations like Dover's Powder, that mixture of ipecacuanha and powdered opium originally prescribed for gout by Dr Thomas Dover. . . . An expanding variety of commercial preparations began to come on the market at mid-century. They were typified by the chlorodynes — Collis Browne's, Towle's and Freeman's. The children's opiates like Godfrey's Cordial and Dalby's Carminative were long established [and used by working mothers to keep their children quiet while they went out to work]. They were everywhere to be bought. There were local preparations, too, like Kendal Black Drop, popularly supposed to be four times the strength of laudanum — and well known outside its locality because Coleridge used it. Poppy head tea in the Fens, "sleepy bear" in the Crickhowell area, Nepenthe, Owbridge's Lung tonic, Battley's sedative solution — popular remedies, patent medicines and the opium preparations of the textbooks were all available.

There were no restrictions on opium use in England until 1868 when the first Pharmacy Act became law. Meanwhile an active anti-opium movement grew in Britain, concerned not so much about the unrestricted use of the drug in Britain but about the way in which Britain encouraged the opium trade in China and in India. A year after the Indian Hemp Drugs Commission (1894) published its conclusions, the British Government's Royal Commission on Opium (1895) published its report on the consumption and trade of opium in India. The Royal Commission had been established in response to pressure from the anti-opium movement in England. The British in India had permitted and even encouraged the large-scale cultivation and trade in opium, with exports to China being particularly important. Around 5000 tons of raw opium went to China each year in the 1880s. The British introduction of opium to the Chinese purely for trade purposes was not one of the events of Imperial history in which we can take any pride. Exports of opium also went from India to supply the growing demand for opium in Britain. In addition, the British administration licensed almost 10,000 shops to sell opium to the local people in India, who consumed a total of nearly 400 tons a year in the period under review (1892–1893). This trade generated very significant income to the British administration in India through excise taxes and license fees, indeed the commission concluded:

> . . . the revenue derived from opium is indispensable for carrying on with efficiency the Government of India. Every native witness who advocated the suppression of the opium traffic admitted that if, as consequence of such a step, taxation in some new form must be imposed, popular opinion would be opposed to any change.

It is, thus not surprising that despite calls for prohibition from several native witnesses and from missionaries and other religious figures, the Royal Commission concluded:

> In summing up the general results of our inquiry, we may first consider the arguments by which support has been obtained for the anti-opium movement. It has been widely held that opium is essentially a poison, used only for vicious and baneful indulgence. Judgement on such a question must mainly turn on medical evidence; and it was abundantly provided for the guidance of your Commission. . . . To the unscientific, the effects of that testimony may perhaps be most clearly conveyed by saying that the temperate use of opium in India should be viewed in the same light as the temperate use of alcohol in England. Opium is harmful, harmless or even beneficial, according to the measure of discretion with which it is used.
>
> Opium is used as a stimulant, and it is largely consumed in India for the mitigation of suffering and the prevention or cure of disease. It is the universal household remedy. It is extensively administered to infants, and the practice does not appear to any appreciable extent injurious. The use of opium does not cause insanity. . . .
>
> As the result of a searching inquiry, and upon a deliberate review of the copious evidence submitted to us, we feel bound to express our conviction that the movement in England in favour of active interference on the part of the Imperial Parliament for the suppression of the opium habit in India, has proceeded from an exaggerated impression as to the nature and extent of the evil to be controlled.
>
> (Royal Commission on Opium, 1895)

Expert medical opinion at the time saw opium as an indispensable part of medicine, and was slow to recognize the dangers that its widespread uncontrolled use might pose for society.

Cocaine

In much the same way, when cocaine was first discovered a century ago as the active component in coca leaves many experts extolled its virtues. Following the discovery of the usefulness of cocaine as a local anesthetic by Carl Koller in 1884 it rapidly gained medical acceptance. For a decade or so it was hugely popular not just as an anesthetic but for a multitude of other uses. Ironically one of its popular uses was in the treatment of opium addicts during their withdrawal phase. During the 1880s Sigmund Freud experimented with this and other applications, and he took the drug himself (Byck, 1974). A number of over-the-counter patent medicines became available that contained coca or cocaine, including various coca wines sold as "restoratives" or tonics. In the 1890s the Burroughs Wellcome Company sold cocaine tabloids that were said to:

> . . . impart a clear and silvery tone to the voice. They were easily retained in the mouth while singing and speaking. . . . used by leading singers and public speakers throughout the world.

The New York physician and scholar W. Golden Mortimer (1901) wrote the *History of Coca. The Divine Plant of the Incas.* As a result of his research he came to some firm conclusions:

> As to the value of Coca, there cannot be the slightest doubt. As to its utter harmlessness there can be no question. Even cocaine, against which there has been a cry of perniciousness, is an ally to the physician of inestimable worth, greatly superior — compare it to a drug of recognised potency, not because of any allied qualities — to morphine. The evils from cocaine have arisen from its pernicious use, in unguarded doses, where used hypodermatically or locally for anaesthesia, when an excessive dose has been administered, without estimating the amount of the alkaloid that would be absorbed, and which might result in systemic symptoms. Medicinally employed, cocaine in appropriate dosage is a stimulant that is not only harm-

less, but usually phenomenally beneficial when indicated . . . Coca is not only a substance innocent as is tea or coffee—which are commonly accepted popular necessities—but that Coca is vastly superior to these substances, and more worthy of general use because of its depurative action on the blood, as well as through its property of provoking a chemico-physiological change in the tissues whereby the nerves and muscles are rendered more capable for their work."

By the turn of the century, however, his was a lone voice, the party was over. It had quickly become clear that cocaine was a dangerous drug of addiction. Apart from its continuing value as a local anesthetic there was no further medical use. Some of the medical experts who had pioneered the introduction of cocaine themselves became victims to the cocaine habit; Freud himself continued to take the drug until the late 1880s. Nonmedical use of the drug was rare on either side of the Atlantic, however, for almost another century—when cocaine reemerged in Western societies in the late twentieth century.

Tobacco

The twentieth century has witnessed the rise and fall of tobacco as a psychoactive drug for the mass of the population in the Western world. Cigarette smoking was comparatively rare at the start of the century; sales in the United States averaged one cigarette per adult male per day. But consumption rose rapidly to around 10 per day by 1945, by then almost half of all adult men were regular smokers—consuming an average of 20 cigarettes a day. Smoking among women lagged somewhat behind, becoming common only some 20 years later. Cigarette smoking was heavily advertised throughout the West, and the advertisements included claims for the alleged health benefits of smoking. In nearly every midcentury Hollywood movie the stars seemed to have a cigarette almost permanently in their mouths.

But things began to change when the link between smoking and lung cancer was first reported in the 1950s. Gradually we have moved from viewing tobacco smoking as a harmless form of relaxation to seeing it is a public health problem that should be dealt with by medical treatments. This change happened gradually; it started with the first definitive health warnings from governments in the 1960s. At about that time the tobacco industry acknowledged that there was a problem and responded by promoting low tar, filter cigarettes. The health warnings had some impact—smoking in America has declined from more than 40% of the adult population to about 25%, and the number of ex-smokers has tripled. In the 1990s this shift has accelerated, nicotine is now recognized to be a drug of addiction and the smoking habit is treated, as with other addictions, by medical means—in this case in the form of nicotine replacement therapy to help smokers quit. The antismoking campaign has reached new heights of emotion in which reason seems sometimes to have been abandoned. Thus, although only a few studies have shown that "passive smoking" carries any real risks to health no-smoking rules govern the work place, hospitals, and other public places and in California even smoking in public bars is prohibited. The negative image of smoking is now so powerful that even history is occasionally rewritten. In the spring of 1999, for example, the United States Post Office issued a 33¢ stamp to commemorate the contribution of the famous artist Jackson Pollock to contemporary art. The stamp used a well-known image of the artist crouched over a canvas, first published in *Life* magazine. In the photograph the artist, a chain-smoker, was smoking a cigarette, but this has miraculously been removed from his lips on the stamp! In the Untied States tobacco companies are increasingly seen as among "the worst of civilization's evil empires." In 1998, the companies agreed to pay more than $200 billion in settlement of present and future claims by smokers whose health has suffered as a result of consuming their products. The final stage in the downfall of tobacco in the United States may come if the FDA gets its wish to be able to regulate nicotine as a dangerous drug.

What Are the Likely Outcomes in
The Twenty-First Century?

The case for relaxing current regulations to make cannabis and cannabis-based medicines more widely available to patients who want them seems overwhelming. Opinion polls on both sides of the Atlantic, and in the United States, and the voters in state elections, have said to governments that this is what they want. There is substantial support for such moves from the medical profession, which is becoming increasingly exasperated at having to connive with patients to recommend an illicit medicine. Whether the voters in the eight American states who asked that medical marijuana be made immediately available will win the day remains to be seen. But it seems increasingly likely that genuine well-controlled clinical trials will take place both in Europe and in the United States in the near future. If such trials yield positive data it would be hard for any government to resist an application for the approval of cannabis or a cannabinoid as a medicine, if this were made through the usual channels for drug approval. The argument that approval of the medical use of cannabis would be tantamount to encouraging the legalization of the drug for all purposes is clearly specious, and is no justification for withholding an effective medicine from patients who need it.

In terms of the debate on recreational use it is more difficult to see a way forward. Too often in the past marijuana has been equated with morality and the debate about its use portrayed as one of good versus evil. Marijuana has been linked with the pursuit of pleasure and with idleness rather than the work ethic. It has rarely been regarded simply as a substance with effects and side effects. Most scientists are now persuaded that the grave short-term risks that cannabis was said to pose to physical and mental health were grossly exaggerated. But the potential long-term health risks of smoked marijuana will remain impossible to assess until data have been amassed over much longer periods of exposure than are available currently. Taking the analogy of cigarette smoking, lung cancer rates among different age groups in the population continued to rise for more than 50 years after cigarette smoking became common among young people. Modern research has also made it clear that the notion that cannabis was not a drug of addiction was false. Some people — perhaps as many as 10% of regular users — will become dependent on the drug. It is not as strongly addictive as cocaine or heroin or even nicotine, but to deny that it has any such liability is no longer reasonable. The recreational use of cannabis has less adverse impact on public health than either tobacco or alcohol, but would this remain the case if the use of the drug became more widespread? At the moment less than 5% of the adult population in the West use cannabis on a daily basis, while around one-third are daily cigarette smokers and more than two-thirds regular users of alcohol. There seems little doubt that an increased use of cannabis would bring an increased public health impact.

Treating the recreational use of cannabis as a crime, however, seems to be both unnecessarily harsh and as a policy very unsuccessful in limiting the use of the drug. In the United States in 1998, 695,000 people were arrested for cannabis offenses, in Britain in 1997 there were 86,086 cannabis offenses. In both countries policing cannabis accounted for nearly 80% of police time on all drug offenses. Although many of those arrested were let off with a police caution (almost two-thirds in Britain) and not prosecuted, they still retained a criminal record for a very minor offense. Many young people have their careers wrecked by expulsion from their schools or colleges for cannabis-related offenses. Others suffer prison sentences, thereby gaining access to a criminal world, which teaches them a great deal more about drugs than they had known before. Surely the laws on drugs should exist to protect society from the ill effects that drug use may cause, not to protect the individual citizen from the folly of their ways. The original prohibition of cannabis was based on false claims that its consumption would lead to criminal behavior, but this proved to be untrue and this rationale no longer exists. Many of our current problems with cannabis stem from the hasty manner in which the drug was classified as a Schedule I narcotic in the 1920s and 1930s and how this classification was not changed when the opportunity later arose. Leo Hollister, a respected academic expert on psychoactive drugs, summed up this point in his testimony to the United States House of Representatives Ways and Means Committee Hearings on the Controlled Substances Act in July 1970:

> I have been unable to find any scientific colleague who agrees that the scheduling of drugs in the proposed legislation makes any sense, nor have I been able to find anyone who was consulted about the proposed schedules. The unfortunate scheduling which groups together such diverse drugs as

heroin, LSD and marihuana perpetuates a fallacy long apparent to our youth. These drugs are not equivalent in pharmacological effects or in the degree of danger they represent to individuals and to society. On the other hand, the specious criterion of medical use places amphetamines in a much lesser category, which the facts do not support. If such scheduling of drugs is retained in the legislation which is ultimately passed, the law will become a laughing stock.

The scheduling was retained in the law and the young have largely disregarded it.

Would it not be better to devote police resources to tackling the abuse of more dangerous drugs? As more and more of the young people who experimented with cannabis in the 1960s and 1970s become parents themselves, attitudes to the recreational use of the drug are likely to become more and more tolerant. In advertising and in pop music the widespread use of cannabis as the "third drug" (alcohol and nicotine beeing the other two) is increasingly acknowledged and has almost attained the chic appeal that cigarette smoking used to have 60 years ago.

On the other hand, can we imagine a society in which cannabis was legally and freely available? One can argue that we have managed to control the availability of alcohol and tobacco, and by taxing these commodities and controlling their price in having some control on the levels of consumption. But are we ready to see commercial companies (perhaps the bruised and battered tobacco companies rising like phoenixes from the ashes) advertising cannabis to consumers? In the alcohol industry advertising targets the high consumer — would this not also happen with cannabis? Would legalisation not inevitably lead to a rise in the number of people using cannabis and probably a rise in their levels of consumption? Would the widespread commercial availability of high-THC cannabis not make it more likely that the problem of cannabis dependence would inevitably increase? These are all imponderables. Perhaps the compromise that we are most likely to reach in the foreseeable future would be something resembling the Dutch experiment — a grudging acceptance that cannabis has become part of our culture, but falling short of full legalisation.

Meanwhile, in political terms, the topic of cannabis is taboo, if it is ever raised the same old warnings about its dangers are reiterated. However, as the *Lancet* editorial of November 11,1995 stated:

> Cannabis has become a political football, and one that governments continually duck. Like footballs, however, it bounces back. Sooner or later politicians will have to stop running scared and address the evidence: cannabis per se is not a hazard to society but driving it further underground may well be.

References

Abel, E. L. *Marihuana.The First Twelve Thousand Years*. New York: Plenum Press, 1943 (Reprinted 1980).

Abood, M. E., and B. R. Martin. Neurobiology of marijuana abuse. *Trends Pharmacol Science* 13:201–206, 1992.

Abrahamov, A., A. Abrahamov, and R. Mechoulam. An efficient new cannabinoid antiemetic in pediatric oncology. *Life Sci* 56:2097–2102, 1995.

Adams, R. Marihuana. *Harvey Lectures* 37:168–197, 1941–42.

Adams, I. B., and B. R. Martin. Cannabis: Pharmacology and toxicology in animals and humans. *Addiction* 91:1585–1614, 1996.

Adler, M. W., and Geller, E. B. Ocular effects of cannabinoids. In *Cannabinoids as Therapeutic Agents*, R. Mechoulam, ed, pp 51–70. Boca Raton: CRC Press, 1986.

Advisory Committee on Drug Dependence. *Cannabis*. London: Her Majesty's Stationery Office, 1969.

Agurell, S., M. Halldin, J-E. Lindgren, A. Ohlsson, M. Widman, H. Gillepie, and L. Hollister. Pharmacokinetics and metabolism of Δ^1-tetrahydrocannabinol and other cannabinoids with emphasis on man. *Pharmacol Rev* 38:21–38, 1986.

D'Ambra, T. E., et al. C-attached aminoalkylindoles: Potent cannabinoid mimetics. *Bioorg Med Chem Lett* 6:17–22, 1996.

American Medical Association. Report of the Council on Scientific Affairs to AMA House of Delegates on *Medical Marijuana*. CSA Report I-97, 1997.

American Psychiatric Association. Diagnostic and Statistical Manual of Mental Disorders (DSM-IV), Fourth Edition. Washington D.C., 1994.

Andreasson, S., P. Allebeck, A. Engstrom, and U. Rydberg. Cannabis and schizophrenia. A longitudinal study of Swedish conscripts. *Lancet* ii:1483–1485, 1987.

Andreasson, S., P. Allebeck, and U. Rydberg. Schizophrenia in users and non-users of cannabis. *Acta Psychiatr Scand* 79:505–510, 1989.

Anslinger, H., and C. R. Cooper. Marihuana: Assassin of youth. *American Magazine*, July:150, 1937.

Atha, M. J., and S. Blanchard, Regular Users. Self-Reported Consumption Patterns and Attitudes Towards Drugs Among 1333 Regular Cannabis Users. 1997 *Independent Drug Monitoring Unit*, Wigan WN2 3ZZ, UK.

Axelrod, J., and C. Felder. Cannabinoid receptors and their endogenous agonist anandamide. *Neurochemical Res* 23:575–581, 1998.

Baddeley, A. The fractionation of working memory. *Proc Natl Acad Sci USA* 93:13468–13472, 1996.

Beal, J. E., et al. Dronabinol as a treatment for anorexia associated with weight loss in patients with AIDS. *J Pain Symptom Management* 10:89–97, 1995.

Bell, R., H. Wechsler, and L. D. Johnston. Correlates of college student marijuana use: results of a US National Survey. *Addiction* 92:571–581, 1997.

Berke, J., and C. Hernton. *The Cannabis Experience*. Aylesbury, UK: Hazell Watson & Viney, London: Quartet Books Ltd, 1974 (reprinted 1977).

Berridge, V., and G. Edwards. *Opium and the People*. Allen Lane, London: St Martin's Press, 1981 (reprinted London: Free Association Books Ltd, 1999).

Bonnie, R. J., and C. H. Whitbread. *The Marihuana Conviction. A History of Marihuana Prohibition in the Unites States*. Charlottesville, VA: University Press of Virginia, 1974.

Braude, M. C. Toxicology of cannabinoids. In *Cannabis and its Derivatives*, W. M. Paton and J. Crown, eds, pp 89–99. Oxford: Oxford University Press, 1972.

Bridge, J. A. Sir William Brooke O'Shaugnessy, M.D., F.R.S., F.R.C.S., F.S.A.: A biographical appreciation by an electrical engineer. *Notes and Records of the Royal Society London* 52:103–120, 1998.

British Medical Association, *Therapeutic Uses of Cannabis*. UK: Harwood Academic Publishers, 1997.

Byck, R., ed. *Sigmund Freud. The Cocaine Papers* (reprints of Freud's writings on cocaine). Stonehill, New York, 1974.

Canadian Government Commission of Inquiry into the non-Medical Use of Drugs. *Cannabis*. Ottawa: Information Canada, 1970.

Chan, P. C., R. C. Sills, A. G. Braun, J. K. Haseman, and J. R. Bucher. Toxicity and carcinogenicity of delta-9-tetrahydrocannabinol in Fischer rats and B6C3F1 mice. *Fundam Appl Toxicol* 30:109–117, 1996.

Chopra, I. C., and R. N. Chopra. The use of cannabis drugs in India. *Bull Narc* January:4–29, 1957.

Clarke, R. C. *Marijuana Botany*. Berkeley, CA: Ronin Publishing, 1981.

Clifford, D. B. Tetrahydrocannabinol for tremor in multiple sclerosis. *Ann Neurol* 13:669–671, 1983.

Comitas, L. Cannabis and work in Jamaica: a refutation of the amotivational syndrome. *Ann N Y Acad Sci* 282:24–32, 1976.

Consroe, P. F., and Snider, R. Therapeutic potential of cannabinoids in neurological disorders. In *Cannabinoids as Therapeutic Agents*, R. Mechoulam, ed, pp 21–49. Boca Raton: CRC Press, 1986.

Consroe, P., R. Musty, J. Rein, W. Tillery, and R. Pertwee. The perceived effects of smoked cannabis on patients with multiple sclerosis. *Eur Neurol* 38:44–48, 1996.

De Fonseca, F. R., M.R.A. Carrera, M. Navarro, G. F. Koob, and F. Weiss. Activation of corticotropin-releasing factor in the limbic system during cannabinoid withdrawal. *Science* 276:2050–2054, 1997.

Devane, W. A., A. Dysarz, M. R. Johnson, L. S. Melvin, and A. Howlett. Determination and characterization of a cannabinoid receptor in rat brain. *Mol Pharmacol* 34:605–613, 1988.

Devane, W. A., L. Hanus, A. Breuer, R. G. Pertwee, L. A. Stevenson, G. Griffin, D. Gibson, A. Mandelbaum, A. Etinger, and R. Mechoulam. Isolation and structure of a brain constituent that binds to the cannabinoid receptor. *Science* 258:1946–1949, 1992.

Di Marzo,V., D. Melck, T. Bisogno, and L. De Petrocellis. Endocannabinoids: Endogenous cannabinoid receptor ligands with neuromodulatory function. *Trends Neurosci* 21:521–528, 1998.

Dixon, W. E. The pharmacology of cannabis indica. *BMJ* Nov 11:1354–1357, 1899.

Doll, R., R. Peto, K. Wheatley, R. Gray, and I. Sutherland. Mortality in relation to smoking: 40 years' observations on male British doctors. *BMJ* 309:901–910, 1994.

Duane Sofia, R. Cannabis: Structure-activity relationships. In *Handbook of Psychopharmacology* Vol. 12 L. L. Iversen, S. D. Iversen, and S. H. Snyder, eds, pp 319–371. New York: Plenum Press, 1978.

Emrich, H. M., F. M. Leweke, and U. Schneider. Towards a cannabinoid hypothesis of schizophrenia: Cognitive impairments due to dysregulation of the endogenous cannabinoid system. *Pharmacol Biochem Behav* 56:8030–8080, 1997.

Engelsman, E. L. Dutch policy on the management of drug-related problems. *Br J Addiction* 84:211–218, 1989.

Felder, C. C., and M. Glass. Cannabinoid receptors and their endogenous agonists. *Annu Rev Pharmacol Toxicol* 38:179–200, 1998.

Fields, H. L., and Meng, I. D. Watching the pot boil: Selective antagonists of the two cannabinoid receptors unveil distinct but synergistic peripheral analgesic activities for endogenous cannabinoids. *Nat Med* 4:1008–1009, 1998.

Fletcher, J. M., Page, J. B., Francis, D. J., Copeland, K., et al. Cognitive correlates of long-term cannabis use in Costa Rican men. *Archiv Gen Psychiatry* 53:1051–1057, 1996.

French.E. D., K. Dillon, and X. Wu. Cannabinoids excite dopamine neurons in the ventral tegmentum and substantia nigra. *NeuroReport* 8:649–652, 1997.

Fried, P. A. Prenatal exposure to tobacco and marijuana: Effects during pregnancy, infancy, and early childhood. *Clin Obstet Gynecol* 36:319–337, 1993.

Ghodse, H. When too much caution can be harmful. *Addiction* 91:764–766, 1996.

Giuffrida, A., L. H. Parsons, T. M. Kerr, F. R. de Fonseca, M. Navarro, and D. Piomelli. Dopamine activation of endogenous cannabinoid signalling in dorsal striatum. *Nat Neurosci* 2:358–363, 1999.

Goode, E. *The Marijuana Smokers*. New York: Basic Books, Inc, 1970.

Grinspoon, L., and J. B. Bakalar. *Marihuana, the Forbidden Medicine*. New Haven, CT: Yale University Press, 1993 (Revised and expanded edition 1997).

Grinspoon, L., and J. B. Bakalar. Marihuana as medicine. A plea for reconsideration. *J.Am. Med.Assoc,* 273:1875–1876, 1995.

Grufferman, S., H. H. Wang, E. R. DeLong, S.Y.S. Kimm, E. S .Delzell, and J. M. Falletta. Environmental factors in the etiology of rhabdomyosarcoma in childhod. *J Nat Cancer Inst* 68:107–113, 1982.

Grufferman, S., et al. Parent's use of cocaine and marijuana and increased risk of rhabdomyosarcoma in their children. *Cancer Causes Control* 4:217–224, 1993.

Hall, W., and N. Solowij. Long-term cannabis use and mental health. *Brit J Psychiatry* 171:107–108, 1997.

Hall, W., N. Solowij, and J. Lemon. *The Health and Psychological Consequences of Cannabis Use* Monograph Series No 25, National Drug Strategy. Canberra: Australian Government Publishing Service, 1994.

Hampson, A. J., M. Grimaldi, and J. Axelrod. Cannabidiol and delta-9-tetrahydrocannabinol are neuroprotective antioxidants. *Proc Natl Acad Sci USA* 95:8268–8273, 1998.

Herkenham, M., A. B. Lynn, M. R. Johnson, L. S. Melvin, B. R. de Costa, and K. C. Rice. Characterization and localization of cannabinoid receptors in rat brain: A quantitative *in vitro* autoradiographic study. *J Neurosci* 11:563–583, 1991.

Herer, J. *The Emperor Wears No Clothes*. Newcastle upon Tyne, UK: Green Plant Co, 1993.

Himmelstein, J. L. The Strange Career of Marihuana. *Contributions in Political Science,* No 94. Westport, CT: Greenwood Press, 1978.

Hollister, L. E. Health aspects of cannabis. *Pharmacol Rev* 38:1–20, 1986.

Hollister, L. E. Marijuana and immunity. *J Psychoactive Drugs* 24:150–164, 1992.

Hollister, L. E. Health aspects of cannabis: Revisited. *Int J Neuropsychopharmacol* 1:71–80, 1998.

House of Lords, Select Committee on Science and Technology, Cannabis — The Scientific and Medical Evidence, London: The Stationery Office, 1998.

Huestis, M. A., A. H. Sampson, B. J. Holicky, J. E. Henningfield, and E. J. Cone. Characterization of the absorption phase of marijuana smoking. *Clin Pharmacol Ther* 52:31–41, 1992.

Indian Hemp Drugs Commission, *Report.* Simla, India: Government Central Printing Office, 1894.

Institute of Medicine. *Marijuana and Medicine.* J. E. Joy, J. Watson, Jr., and J. A. Benson, Jr., eds Washington, D.C.: National Academy Press, 1999.

Jahr, G.H.G. *New Homeopathic Pharmacapoeia and Posology or the Preparation of Homeopathic Medicines,* p.137. Philadelphia: J.Dobson, 1842.

Kandel, D. B., and M. Davies. High school students who use crack and other drugs. *Arch Gen Psychiatry* 53:71–80, 1996.

Kaslow, R. A., W. C. Blackwelder, D. G. Ostrow, D. Yerg, J. Palenicek, A. H. Coulson, and R. O. Validiserri. No evidence for a role of alcohol or other psychoactive drugs in accelerating immunodeficiency in HIV-1-positive individuals. *J Am Med Assoc* 261:3424–3429, 1989.

Kolodny, R. C., W. H. Masters, R. M. Kolodner, and G. Toro. Depression of plasma testosterone levels after chronic intensive marihuana use. *N Engl J Med* 290:872–874, 1974.

Kuster, J. E., J. I. Stevenson, S. J. Ward, T. E. D'Ambra, and D. A. Haycock. Aminoalkylindole binding in rat cerebellum: Selective displacement by natural and synthetic cannabinoids. *J Pharmacol Exp Ther* 264:1352–1363, 1993.

Ledent, C, et al. Unresponsiveness to cannabinoids and reduced addictive effects of opiates in CB$_1$ receptor knockout mice. *Science* 283:401, 1999.

Lemberger, L. Clinical evaluation of cannabinoids in the treatment of disease. In *Marihuana '84,* D. J. Harvey, W. Paton, and G. Nahas, eds., pp 673–680. Oxford: IRL Press Ltd, 1985.

Levine, J. D., N. C. Gordon, J. C. Bornstein, and H. L. Fields. Role of pain in placebo analgesia. *Proc Natl Acad Sci USA* 76:3528–3531, 1979.

Levitt, M., C. Faiman, R. Hawks, and A. Wilson. Randomized double-blind comparison of delta-9-tetrahydrocannabinol (THC) and marijuana as chemotherapy antiemetics. *Proc Am Soc Clin Oncol* 3:91, 1984.

Lewin, L. *PHANTASTICA. Narcotic and Stimulating Drugs their Use and Abuse.* London: Routledge & Kegan Paul, 1931.

Linzen, D. H., P. M. Dingemans, and M. E. Lenoir. Cannabis abuse and the course of recent onset schizophrenia disorders. *Arch Gen Psychiaty* 51:273–279, 1994.

Liu, B-Q., et al. Emerging tobacco hazards in China: 1. Retrospective proportional mortality study of one million deaths. *Br Med J* 317:1411–1422, 1998.

Ludlow, F. H. *The Hasheesh Eater: Being Passages from the Life of a Pyathagorean.* New York: Harper Bros, 1857.

MacCoun, R., and P. Reuter. Interpreting Dutch cannabis policy: Reasoning by analogy in the legalization debate. *Science* 278:47–52, 1997.

Makriyannis, A., and Rapaka, R. The molecular basis of cannabinoid activity. *Life Sci* 47:2173–2184, 1990.

Marshall, C. R. The active principle of Indian Hemp: a preliminary communication. *Lancet* i:235–238, 1897.

Martin, B. R. Characterization of the antinociceptive activity of Δ^9-tetrahydrocannabinol in mice. In *Marihuana '84*, D. J. Harvey, ed, pp 685–692. Oxford: IRL Press, 1985.

Martyn, C. N., L. S. Illis, and J. Thorn. Nabilone in the treatment of multiple sclerosis. *Lancet* i:579, 1995.

Matsuda, L. A., S. J. Lolait, M. J. Brownstein, A. C. Young, and T. I. Bonner. Structure of a cannabinoid receptor and functional expression of the cloned cDNA. *Nature* 346:561–564, 1990.

Matthew, R. J., W. H. Wilson, R. E. Coleman, T. G. Turkington, and T. R. DeGrado. Marijuana intoxication and brain activation in marijuana smokers. *Life Sciences* 60:2075–2089, 1997.

Mayor La Guardia's Committee on Marihuana. *The Marihuana Problem in the City of New York*, J. Cattell Press, Lancaster, PA: 1944.

Mechoulam, R. Marihuana chemistry. *Science* 168:1159–1163, 1970.

Mechoulam, R., A. Shani, B.Yagnitinsky, Z. Ben-Zvi, P. Braun, and Y. Gaoni. Some aspects of cannabinoid chemistry. In *The Botany and Chemistry of Cannabis*, C.R.B. Joyce and S. H. Curry, eds, pp 93–117. Churchill, London: CIBA Foundation Conference, 1970.

Melges, F. T., J. R. Tinklenberg, L. E. Hollister, and H. K. Gillespie. Marihuana and the temporal span of awareness. *Arch Gen.Psychiatry* 24:564–567, 1971.

Merritt, J. C., J. L. Olsen, P. C. Alexander, A. L. Anduze, and S. S. Gelbart. Effect of marijuana on intraocular and blood pressure in glaucoma. *Ophthalmology* 87:222–228, 1980.

Moon, J. B. Sir William Brooke O'Shaugnessy—the foundations of fluid therapy and the Indian telegraph service. *N Engl J Med* 276:283–284, 1967.

Mortimer, W. G. *History of Coca. The Divine Plant of the Incas.* New York: J.H.Vail & Co, 1901.

Nahas, G. *Marihuana—Deceptive Weed.* New York: Raven Press, 1973.

Nahas, G. *Keep Off the Grass.* New York: Reader's Digest Press, 1976.

Nahas, G., G. Suciv-Foca, J-P. Armand, and A. Morishima. Inhibition of cellular mediated immunity in marihuana smokers. *Science* 183:419–420, 1974.

National Commission on Marihuana and Drug Abuse, First Report. *Marihuana: A Signal of Misunderstanding*, 1972, and Second Report, *Drug Use in America: Problem in Perspective*, 1973, Washington D.C.: US Government Printing Office.

National Institutes of Health. *Report on the Medical Uses of Marijuana*. Bethesda, MD, 1997.

Navarro, M. et al. Acute administration of the CB_1 cannabinoid receptor antagonist SR141716A induces anxiety-like response in the rat. *NeuroReport* 8:491–496, 1997.

Negrete, J. C., W. P. Knapp, D. E. Douglas, and W. B. Smith. Cannabis affects the severity of schizophrenic symptoms: Results of a clinical survey. *Psychol Med* 16:515–520, 1986.

O'Shaugnessey, W. B. On the preparation of the Indian hemp, or gunjah (cannabis indica) and their effects on the animal system in health and their utility in the treatment of tetanus and other convulsive disorders. *Trans Med Phys Soc Calcutta* 8:421–461, 1842.

Ossebaard, H. C. Netherlands' cannabis policy. *Lancet* 347:767–768, 1996.

Pate, D. W., K. Jarvinen, A. Urtti, V. Mahadevan, and T. Jarvinen. Effect of the CB1 receptor antagonist, SR141716A on cannabinoid-induced ocular hypotension in normotensive rabbits. *Life Sci* 63:2181–2188, 1998.

Peralta, V., and M. J. Cuesta. Influence of cannabis abuse on schizophrenic psychopathology. *Acta Psychiatr Scand* 85:127–130, 1992.

Pério, A., M. Rimaldi-Carmona, J. Maruani, F. Barth, G. Le fur, and P. Soubrié. Central mediation of the cannabinoid cue: Activity of a selective CB1 antagonist, SR 141716A. *Behav Pharmacol* 7:65–71, 1996.

Pertwee, R. G. Tolerance to and dependence on psychotropic cannabinoids. In *The Biological Basis of Drug Tolerance* J. Pratt, ed, pp 231–265. London: Academic Press Ltd, 1991.

Pertwee, R., ed. *Cannabinoid Receptors*. London: Academic Press, 1995.

Peto, R. Influence of dose and duration of smoking on lung cancer rates. In *Tobacco: A Major International Health Hazard* D. G. Zaridze and R. Peto, eds, pp 23–33. International Agency for Research on Cancer, Publication No 74, Lyon, France, 1986.

Peto, R., A. D. Lopez, J. Boreham, M. Thun, C. Heath, Jr., and R. Doll. Mortality from smoking worldwide. *Br Med Bull* 52:12–21, 1996.

Piomelli, D., M. Beltramo, A. Giuffrida, and N. Stella. Endogenous cannabinoid signalling, *Neurobiol Dis* 5:462–473, 1998.

Plasse, T. F., R. W. Gorter, S. H. Krasnow, M. Lane, K. V. Shepard, and R. G. Wadleigh. Recent clinical experience with dronabinol. *Pharmacol Biochem Behav* 40:695–700, 1991.

Polen, M. R., S. Sidney, I. S. Tekawa, M. Sadler, and G. D. Friedman. Health care use by frequent marijuana smokers who do not smoke tobacco. *West J Med* 158:596–601, 1993.

Pope, H. G. Jr., Ionescu-Pioggia, M., Aizley, H. G., and Varma, D. K. Drug use and life style among college undergraduates in 1989: a comparison with 1969 and 1978. *Am J.Psychiatry*, 147:998–1001, 1990.

Randall, R. C., ed. *Muscle Spasm, Pain and Marijuana Therapy*. Washington DC: Galen Press, 1991.

Reynolds, J. R. On the therapeutic uses and toxic effects of cannabis indica. *Lancet* i:637–638, 1890.

Robinson, R. *The Great Book of Hemp*. Rochester, VT: Park Street Press, 1996.

Robison, L. L., J. D. Buckley, A. E. Daigle, R. Wells, D. Benjamin, D. C. Arthur, and G. D. Hammond. Maternal drug use and risk of childhood non-lymphoblastic leukemia among offspring. *Cancer* 63:1904–1911, 1989.

Rohr, J. M., S. W. Skowlund, and T. C. Martin. Withdrawal sequelae to cannabis use. *Int J Addic* 24:627–631, 1989.

Rosenthal, E., D. Gieringer, and T. Mikuriya. *Marijuana Medical Handbook. A Guide to Therapeutic Use*. Oakland: Quick American Archives, 1997.

Royal Commission on Opium. *Final Report*. London: Her Majesty's Stationery Office, 1895.

Rubin, V., ed. *Cannabis and Culture*. The Hague: Mouton Publishers, 1975.

Russo, E. Cannabis for migraine treatment: the once and future prescription? An historical and scientific review. *Pain* 76:3–8, 1998.

Schwartz, R. H., E. A. Voth, and M. J. Sheridan. Marijuana to prevent nausea and vomiting in cancer patients: A survey of clinical oncologists. *South Med J* 90:167–172, 1997.

Sidney, S., J. E. Beck, I. S. Tekawa, C. P. Quesenberry, and G. D. Friedman. Marijuana use and mortality. *Am J Public Health* 87:585–590, 1997.

Smith, F. L, D. Cichewicz, Z. L. Martin, and S. P. Welch The enhancement of morphine antinociception in mice by delta-9-tetrahydrocannabinol. *Pharmacol Biochem Behav* 60:559–566, 1998.

Smith, T., and H. Smith. Process for preparing cannabine or hemp resin. *Pharmaceutical J* 6:171–173, 1846.

Snyder, S. H. Uses of Marijuana. New York: Oxford University Press, 1971.

Solowij, N. *Cannabis and Cognitive Functioning*. Cambridge, UK: Cambridge University Press, 1998.

Tanda, G., F. E. Pontieri, and G. Di Chiara. Cannabinoid and heroin activation of mesolimbic dopamine transmission by a common μ_1 opioid receptor mechanism. *Science* 276:2048–2050, 1997.

Tashkin, D. P. Marihuana and the lung. In *Marihuana and Medicine*, pp 279–288 G. Nahas, S. Agurell, K. M. Sutin, and D. J. Harvey, eds. New York: Humana Press, 1999.

Terranova, J. P., et al. Improvement of memory in rodents by the selective CB1 cannabinoid receptor antaognist, R141716. *Psychopharmacology* 126:165–172, 1996.

Thomas, H. Psychiatric symptoms in cannabis users. *Br J Psychiatry* 163:141–149, 1993.

Thornicroft, G. Cannabis and psychosis. Is there epidemiological evidence for an association? *Br J Psychiatry* 157:25–33, 1990.

Todd, A. R. Hashish. *Experientia* 2:55–60, 1946.

Torbjörn, U., C. Järbe, and B. G. Henriksson. Discriminative response control produced with hashish, tetrahydrocannabinols, and other drugs. *Psychopharmacology* 40:1–16, 1974.

Tunving, K., T. Lundquist, and G. D. Eriksson, A Way Out of the Fog: An outpatient program for cannabis users. In *Marijuana: An International Research Report*, G. Chester, P. Consroe, and R. Musty, eds. Canberra: Australian Government Publishing Service, 1995.

Twycross, R. G. Opioids. In *Textbook of Pain*, P. D. Wall, and Melzack, eds, pp 943–953, London: Churchill Livingstone, 1994.

Van Amsterdam, J.G.C., J. W. van der Laan, and J. L. Slangen. Residual effects of prolonged heavy cannabis use. *National Institute of Public Health and the Environment*, Report No. 318902003. Bilthoven, Netherlands, 1996.

Viniciguerra, V., T. Moore, and E. Brennan. Inhalation marijuana as an antiemetic for cancer chemotherapy. *NY State Med J* 88:525–527, 1988.

Walton, R. P. *Marihuana. America's New Drug Problem*. New York: J.B.Lippincott Co, 1938.

Weil, A. T., N. E. Zinberg, and J. M. Nelsen. Clinical and psychological effects of marihuana in man. *Science* 162:1234–1242, 1968.

WHO. *Cannabis: A Health Perspective and Research Agenda*. Geneva: World Health Organization, 1997.

Wiley, J. L., R. L. Barret, J. Lowe, R. L. Balster, and B. R. Martin. Discriminative stimulus effects of CP-55,940 and structurally dissimilar cannabinoids in rats. *Neuropharmacology* 34:669–676, 1995.

Williams, S. J., J.P.R. Hartley, and J.D.P. Graham. Bronchodilator effects of delta-9-THC administered by aerosol to asthmatic patients. *Thorax* 31:720–723, 1976.

Williams, J. H., N. A. Wellman, and J.N.P. Rawlins. Cannabis use correlates with schizotypy in healthy people. *Addiction* 91:869–87, 1996.

Wisset, R. *A Treatise on Hemp*. London: J. Harding, 1808.

Wood, G. B., and F. Bache. *The Dispensatory of the United States*, p.339. Philadelphia: Lippincott, 1854.

Woolf, P. O. *Marihuana in Latin America. The Threat That It Constitutes*. Washington, D.C.: Linacre Press Inc, 1949.

Wu, T. C., D. P. Tashkin, B. Djaheb, and J. E. Rose. Pulmonary hazards of smoking marijuana as compared with tobacco. N Engl J Med 31:347–351, 1988.

Zimmer, L., and J. P. Morgan. Marijuana Myths, Marijuana Facts. New York: Lindesmith Center, 1997.

Zimmerman, S., and A. M. Zimmerman. Genetic effects of marijuana. Int J Addict 25:19–33, 1990.

Zuardi, A. W., I. Shirakawa, E. Finkelfarb, and I. G. Karniol. Action of cannabidiol on the anxiety and other effects produced by delta-9-THC in normal subjects. Psychopharmacology 76:245–250, 1982.

Zuckerman, B., D. A. Frank, R. Hingson, H. Amdro, S. M. Levenson, H. Kayne, S. Parker, and R.Vinci. Effects of maternal marijuana and cocaine use on fetal growth. N Engl J Med 320:762–768, 1989.

Index

Page numbers followed by *f* and *t* indicate figures and tables, respectively